STUDENT WORKBOOK

college physics

a strategic approach **4e**

knight • jones • field

randall d. knight

California Polytechnic State University, San Luis Obispo

Editor in Chief, Director Physical Science Courseware Portfolio: Jeanne Zalesky
Courseware Portfolio Manager: Darien Estes
Senior Content Producer: Martha Steele
Managing Producer: Kristen Flathman
Courseware Director, Content Development: Jennifer Hart
Senior Analyst, Content Development, Science: Suzanne Olivier
Courseware Editorial Assistant: Kristen Stephens
Rich Media Content Producer: Dustin Hennessey
Full-Service Vendor: Nesbitt Graphics/Cenveo
Compositor: Nesbitt Graphics/Cenveo
Art and Design Director: Mark Ong, Side By Side Studios
Interior/Cover Designer: tani hasegawa
Cover Printer: LSC Communications
Printer: LSC Communications
Manufacturing Buyer: Stacey J. Weinberger/LSC Communications
Product Marketing Manager, Physical Sciences: Elizabeth Bell

Cover Photo Credit: MirageC/Getty

 Pearson

ISBN 10: 0-134-60989-1; ISBN 13: 978-0-134-60989-8

www.pearson.com

11 2019

Table of Contents

Preface

It is highly unlikely that one could learn to play the piano by only reading about it. Similarly, reading physics from a textbook is not the same as doing physics. To develop your ability to do physics, your instructor will assign problems to be solved both for homework and on tests. Unfortunately, it is our experience that jumping right into problem solving after reading and hearing about physics often leads to poor "playing" techniques and an inability to solve problems for which the student has not already been shown the solution (which isn't really "solving" a problem, is it?). Because improving your ability to solve physics problems is one of the major goals of your course, time spent developing techniques that will help you do this is well spent.

Learning physics, as in learning any skill, requires regular practice of the basic techniques. That is what this *Student Workbook* is all about. The workbook consists of exercises that give you an opportunity to practice techniques and strengthen your understanding of concepts presented in the textbook and in class. These exercises are intended to be done on a daily basis, right after the topics have been discussed in class and are still fresh in your mind. Successful completion of the workbook exercises will prepare you to tackle the more quantitative end-of-chapter homework problems in the textbook.

You will find that many of the exercises are *qualitative* rather than *quantitative*. They ask you to draw pictures, interpret graphs, use ratios, write short explanations, or provide other answers that do not involve calculations. A few math-skills exercises will ask you to explore the mathematical relationships and symbols used to quantify physics concepts but do not require a calculator. The purpose of all of these exercises is to help you develop the basic thinking tools you'll later need for quantitative problem solving. It is highly recommended that you do these exercises *before* starting the end-of-chapter problems.

One example from Chapter 4 illustrates the purpose of this *Student Workbook*. In that chapter, you will read about a technique called a "free-body diagram" that is helpful for solving problems involving forces. Sometimes, students mistakenly think that the diagrams are used by the instructor only for teaching purposes and may be abandoned once Newton's laws are fully understood. On the contrary, professional physicists with decades of problem-solving experience still routinely use these diagrams to clarify the problem and set up the solution. Many of the other techniques practiced in the workbook, such as ray diagrams, graphing relationships, sketching field lines and equipotentials, etc., fall in the same category. They are used at all levels of physics, not just as a beginning exercise. And many of these techniques, such as analyzing graphs and exploring multiple representations of a situation, have important uses outside of physics. Time spent practicing these techniques will serve you well in other endeavors.

You will find that the exercises in this workbook are keyed to specific sections of the textbook in order to let you practice the new ideas introduced in that section. You should keep the text beside you as you work and refer to it often. You will usually find Tactics Boxes, figures, or examples in the textbook that are directly relevant to the exercises. When asked to draw figures or diagrams, you should attempt to draw them so that they look much like the figures and diagrams in the textbook.

Because the exercises go with specific sections in the text, you should answer them on the basis of information presented in *just* that section (and prior sections). You may have learned new ideas in Section 7 of a chapter, but you should not use those ideas when answering questions from Section 4. There will be ample opportunity in the Section 7 exercises to use that information there.

You will need a few "tools" to complete the exercises. Many of the exercises will ask you to *color code* your answers by drawing some items in black, others in red, and perhaps yet others in blue. You need to purchase a few colored pencils to do this. The authors highly recommend that you work in pencil, rather than ink, so that you can easily erase. Few are the individuals who make so few mistakes as to be able to work in ink! In addition, you'll find that a small, easily carried six-inch ruler will come in handy for drawings and graphs.

As you work your way through the textbook and this workbook, you will find that physics is a way of *thinking* about how the world works and why things happen as they do. We will primarily be interested in finding relationships, seeking explanations, and developing techniques to make use of these relationships, only secondarily in computing numerical answers. In many ways, the thinking tools developed in this workbook are what the course is all about. If you take the time to do these exercises regularly and to review the answers, in whatever form your instructor provides them, you will be well on your way to success in physics.

To the instructor: The exercises in this workbook can be used in many ways. You can have students work on some of the exercises in class as part of an active-learning strategy. Or you can do the same in recitation sections or laboratories. This approach allows you to discuss the answers immediately, to answer student questions, and to improvise follow-up exercises when needed. Having the students work in small groups (two to four students) is highly recommended.

Alternatively, the exercises can be assigned as homework. The pages are perforated for easy tear-out, and the page breaks are in logical places so that you can assign the sections of a chapter that you would likely cover in one day of class. Exercises should be assigned immediately after presenting the relevant information in class and should be due at the beginning of the next class. Collecting them at the beginning of class, and then going over two or three that are likely to cause difficulty, is an effective means of quickly reviewing major concepts from the previous class and launching a new discussion.

If used as homework, it is *essential* for students to receive *prompt* feedback. Ideally, this would occur by having the exercises graded, with written comments, and returned at the next class meeting. Posting fairly detailed answers on a course website also works. Lack of prompt feedback can negate much of the value of these exercises. Placing similar qualitative/graphical questions on quizzes and exams, and telling students at the beginning of the term that you will do so, encourages students to take the exercises seriously and to check the answers.

Student feedback from end-of-term questionnaires reveals three prevalent attitudes toward the workbook exercises:

   i. They think it is an unreasonable amount of work.
  ii. They agree that the assignments force them to keep up and not get behind.
  iii. They recognize, by the end of the term, that the workbook is a valuable learning tool.

However you choose to use these exercises, they will significantly strengthen your students' conceptual understanding of physics.

Following the workbook exercises are optional Dynamics Worksheets, Momentum Worksheets, and Energy Worksheets for use with end-of-chapter problems in Parts I and II of the textbook. Their use is recommended to help students acquire good problem-solving habits early in the course. If you wish your students to use these, have them make enough photocopies for use throughout the term.

Answers to all workbook exercises are provided as pdf files and can be downloaded from the Instructor Resource Area in Mastering™ Physics as well as from the textbook's Instructor Resource Center at www.pearson.com.

Acknowledgments: The author would like to thank Cenveo Publishing Services for their production of the workbook.

1 Representing Motion

1.1 Motion: A First Look

Exercises 1–3: Draw a motion diagram for each motion described below. Six to eight images are appropriate for most motion diagrams.

1. A car accelerates forward from a stop sign. It eventually reaches a steady speed of 45 mph.

2. A heavy rock is dropped over the edge of a cliff. Draw the motion diagram from the instant the rock is released until the instant it touches the ground.

3. A skier starts *from rest* at the top of a 30° snow-covered slope and steadily speeds up as she skies to the bottom. (Orient your diagram as seen from the *side*. Label the 30° angle.)

1.2 Models and Modeling

Exercises 4–5: Draw a motion diagram for each motion described below.
- Use the particle model to represent the object as a particle.
- Show the object at six or eight positions.
- Number the positions in order, as shown in Figure 1.4 in the text.

4. Alicia is driving at a steady 30 mph when the light ahead turns red. She immediately slows and stops. Show her motion from a few seconds before the light changes until she stops. Draw a vertical dashed line across your motion diagram to show the instant the light changes.

5. Bob is standing on the ground. He throws a ball upward at a 45° angle. The ball lands on a second-story balcony. Show the motion from the instant the ball leaves Bob's hand until the instant it touches the balcony.

Exercises 6–9: For each motion diagram, write a short description of the motion of an object that will match the diagram. Your descriptions should name *specific* objects and be phrased similarly to the descriptions of Exercises 1 to 5 above. Note the axis labels on Exercises 8 and 9.

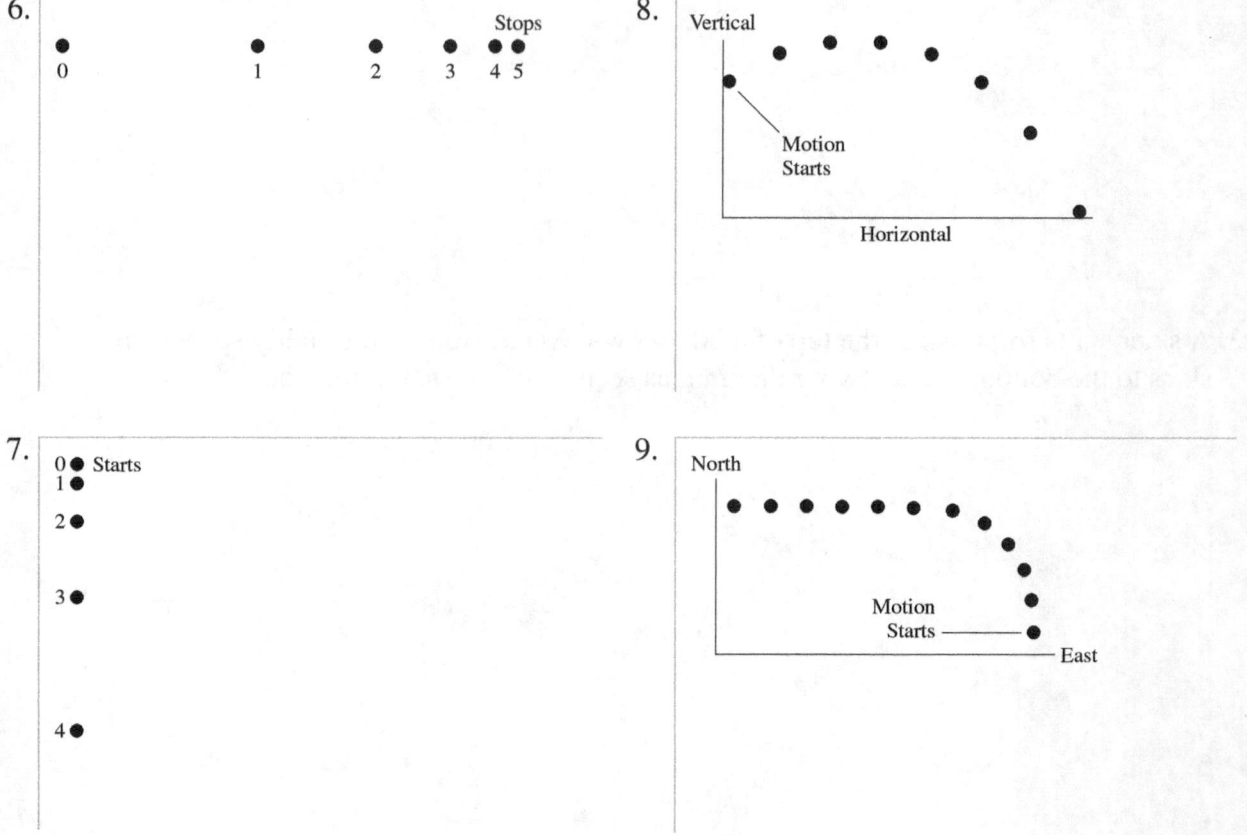

1.3 Position and Time: Putting Numbers on Nature

10. Using the particle model, redraw each of the motion diagrams from Exercises 1 to 3 in the space below. Add a coordinate axis to each drawing and label the initial and final positions. Draw an arrow on your diagram to represent the displacement from the beginning to the end of the motion.

11. In the picture below, Joe starts walking slowly but at constant speed from his house on Main Street to the bus stop 200 m down the street. When he is halfway there, he sees the bus and steadily speeds up until he reaches the bus stop.

    a. Draw a motion diagram in the street of the picture to represent Joe's motion.

    b. Add a coordinate axis below the picture with Joe's house as the origin. Label Joe's initial position at the start of his walk as x_1, his position when he sees the bus as x_2, and his final position when he arrives at the bus stop as x_3. Draw arrows above the motion diagram to represent Joe's displacement from his initial position to his position when he first sees the bus and the displacement from where he sees the bus to the bus stop. Label these displacements Δx_1 and Δx_2, respectively.

    c. Repeat part b in the space below but with the origin of the coordinate axis at the location where Joe starts to speed up.

    d. Do the displacement arrows change when you change the location of the origin?

1.4 Velocity

12. A moth flies a distance of 3 m in only one-third of a second.

  a. What does the ratio 3/(1/3) tell you about the moth's motion? Explain.

  b. What does the ratio (1/3)/3 tell you about the moth's motion?

  c. How far would the moth fly in one-tenth of a second?

  d. How long does the moth take to fly 4 m?

13. a. If someone drives at 25 mph, is it necessary that he or she does so for an hour?

  b. Is it necessary to have a cubic centimeter of gold to say that gold has a density of 19.3 g/cm^3? Explain.

1.5 A Sense of Scale: Significant Figures, Scientific Notation, and Units

14. How many significant figures does each of the following numbers have?

    a. 6.21 _____   e. 0.0621 _____   i. 1.0621 _____

    b. 62.1 _____   f. 0.620 _____   j. 6.21×10^3 _____

    c. 6210 _____   g. 0.62 _____   k. 6.21×10^{-3} _____

    d. 6210.0 _____   h. .62 _____   l. 62.1×10^3 _____

15. Compute the following numbers, applying the significant figure standards adopted for this text.

    a. $33.3 \times 25.4 =$ _____   e. $2.34 \times 3.321 =$ _____

    b. $33.3 - 25.4 =$ _____   f. $(4.32 \times 1.23) - 5.1 =$ _____

    c. $33.3 \div 45.1 =$ _____   g. $33.3^2 =$ _____

    d. $33.3 \times 45.1 =$ _____   h. $\sqrt{33.3} =$ _____

16. Express the following numbers and computed results in scientific notation, paying attention to significant figures.

    a. $9,827 =$ _____   d. $32,014 \times 47 =$ _____

    b. $0.000000550 =$ _____   e. $0.059 \div 2,304 =$ _____

    c. $3,200,000 =$ _____   f. $320 \times 0.050 =$ _____

17. Convert the following to SI units. Work across the line and show all steps in the conversion. Use scientific notation and apply the proper use of significant figures. **Note:** Think carefully about g and h. Pictures may help.

    a. $9.12 \, \mu s = 9.12 \, \mu s \times$

    b. $3.42 \, km = 3.42 \, km \times$

    c. $44 \, cm/ms = 44 \, cm/ms \times$

    d. $80 \, km/hr = 80 \, km/hr \times$

    e. $60 \, mph = 60 \, mph \times$

    f. $8 \, in = 8 \, in \times$

    g. $14 \, in^2 = 14 \, in^2 \times$

    h. $250 \, cm^3 = 250 \, cm^3 \times$

18. Use Table 1.4 and your common sense to assess whether or not the following statements are *reasonable*. Explain. See Examples 1.4 and 1.5 for examples of the type of reasoning needed.

a. Joe is 180 cm tall.

b. I rode my bike to campus at a speed of 50 m/s.

c. A skier reaches the bottom of the hill skiing at a speed of 25 m/s.

d. I can throw a ball at a distance of 2 km.

e. I can throw a ball at a speed of 50 km/hr.

f. Joan's newborn baby has a mass of 33 kg.

g. A hummingbird has a mass of 3.3 g.

1.6 Vectors and Motion: A First Look

19. For the following motion diagrams, draw an arrow to indicate the displacement vector between the initial and final positions.

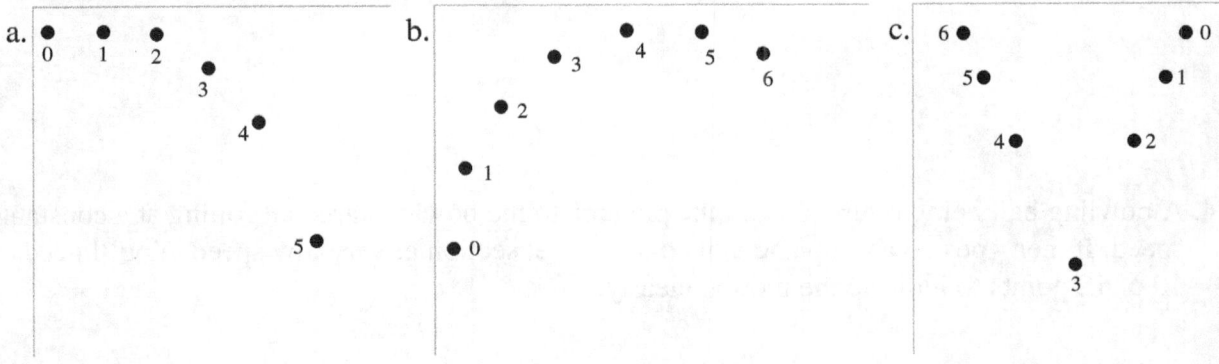

20. In each part of Exercise 19, is the object's displacement equal to the distance the object travels? Explain.

21. Draw and label the vector sum $\vec{A} + \vec{B}$.

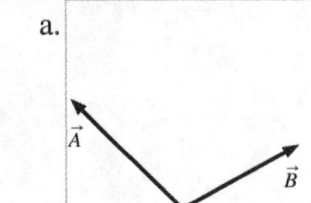   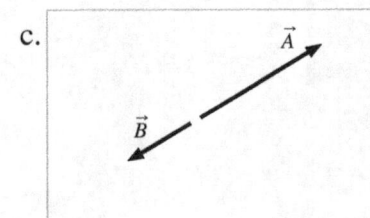

Exercises 22–26: Draw a motion diagram for each motion described below.
- Use the particle model.
- Show and label the *velocity* vectors.

22. Galileo drops a ball from the Leaning Tower of Pisa. Consider the ball's motion from the moment it leaves his hand until a microsecond before it hits the ground. Your diagram should be vertical.

23. A rocket-powered car on a test track accelerates from rest to a high speed, then coasts at constant speed after running out of fuel. Draw a vertical dashed line across your diagram to indicate the point at which the car runs out of fuel.

24. A bowling ball being returned from the pin area to the bowler starts out rolling at a constant speed. It then goes up a ramp and exits onto a level section at very low speed. You'll need 10 or 12 points to indicate the motion clearly.

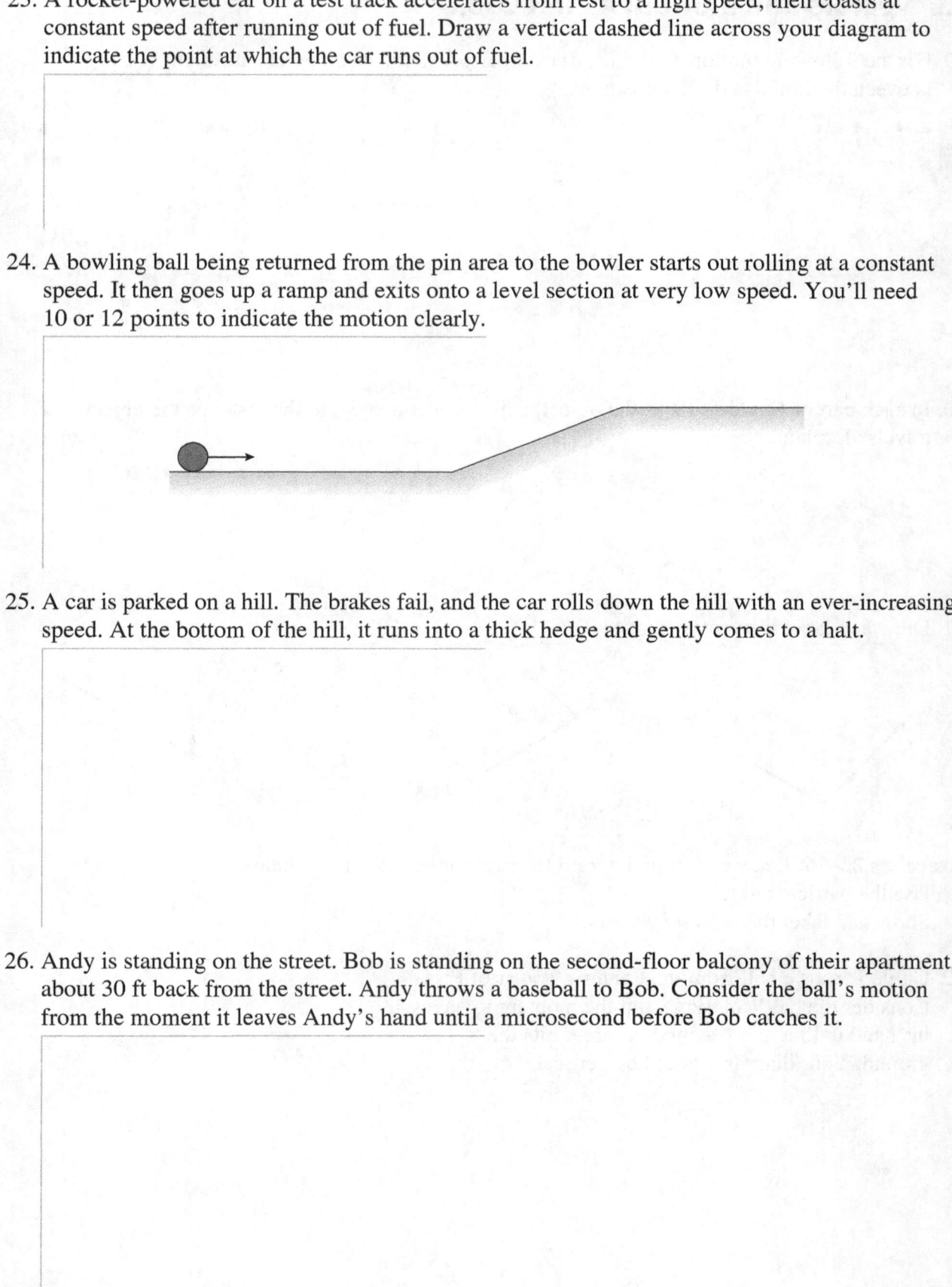

25. A car is parked on a hill. The brakes fail, and the car rolls down the hill with an ever-increasing speed. At the bottom of the hill, it runs into a thick hedge and gently comes to a halt.

26. Andy is standing on the street. Bob is standing on the second-floor balcony of their apartment, about 30 ft back from the street. Andy throws a baseball to Bob. Consider the ball's motion from the moment it leaves Andy's hand until a microsecond before Bob catches it.

2 Motion in One Dimension

2.1 Describing Motion

1. The position-versus-time graph on the right shows the position of an object moving in a straight line for 12 seconds.

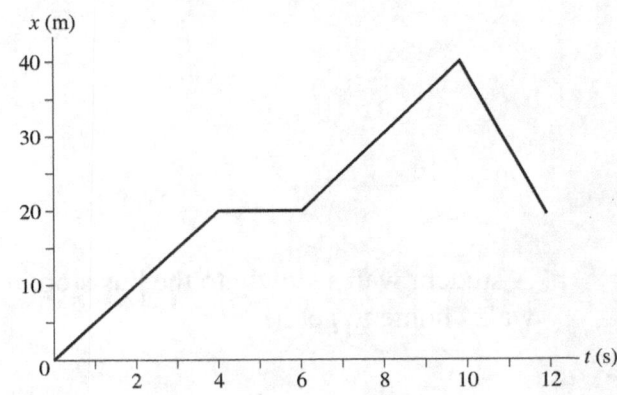

a. What is the position of the object at 2 s, 6 s, and 10 s after the start of the motion?

At 2 s: _____ At 6 s: _____ At 10 s: _____

b. What is the object's velocity during the first 4 s of motion?

c. What is the object's velocity during the interval from $t = 4$ s to $t = 6$ s?

d. What is the object's velocity during the 4 s from $t = 6$ s to $t = 10$ s?

e. What is the object's velocity during the final 2 s from $t = 10$ s to $t = 12$ s?

2. Sketch position-versus-time graphs for the following motions. Include a numerical scale on both axes with units that are *reasonable* for this motion. Some numerical information is given in the problems. For other quantities, make reasonable estimates.

Note: A *sketched* graph is hand-drawn, rather than laid out with a ruler. Even so, a sketch must be neat, accurate, and include axis labels.

a. A student walks to the bus stop, waits for the bus, and then rides to campus. Assume that all the motion is along a straight street.

b. A student walks slowly to the bus stop, realizes he forgot his paper that is due, and *quickly* walks home to get it.

c. The quarterback drops back 10 yards from the line of scrimmage, and then throws a pass 20 yards to a receiver, who catches it and sprints 20 yards to the goal. Draw your graph for the *football*. Think carefully about what the slopes of the lines should be.

3. Interpret the following position-versus-time graphs by writing a very short "story" of what is happening. Be creative! Have characters and situations! Simply saying that "a car moves 100 m to the right" doesn't qualify as a story. Your stories should make *specific reference* to information you obtain from the graphs, such as distances moved or time elapsed.

a. Moving car

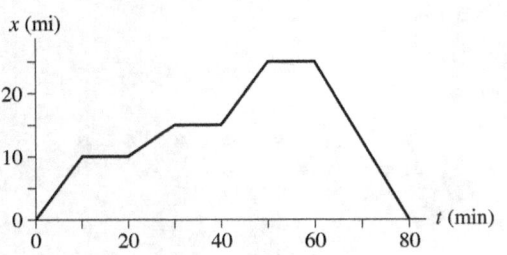

b. Sprinter

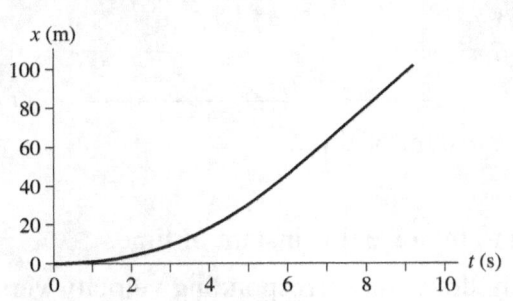

c. Two football players

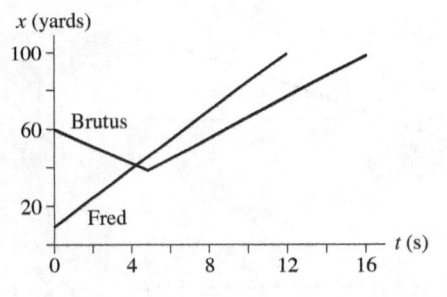

4. The figure shows a position-versus-time graph for the motion of objects A and B that are moving along the same axis.

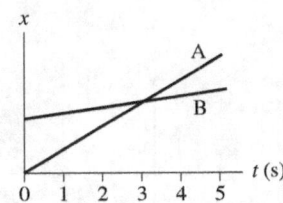

a. At the instant $t = 1$ s, is the speed of A greater than, less than, or equal to the speed of B? Explain.

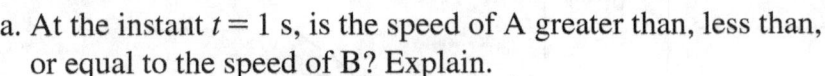

b. Do objects A and B ever have the *same* speed? If so, at what time or times? Explain.

5. Draw both a position-versus-time graph *and* a velocity-versus-time graph for an object that is at rest at $x = 1$ m.

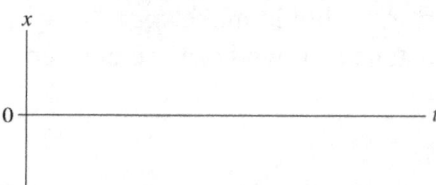 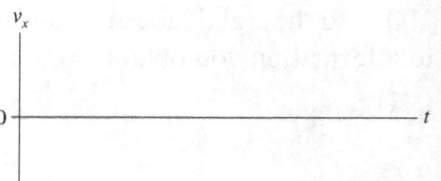

6. The figure shows six frames from the motion diagram of two moving cars, A and B.

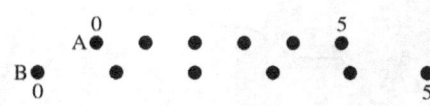

   a. Draw both a position-versus-time graph and a velocity-versus-time graph. Show the motion of *both* cars on each graph. Label them A and B.

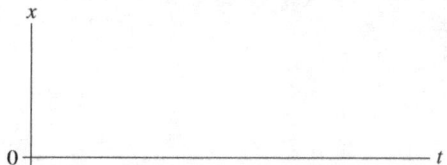 

   b. Do the two cars ever have the same position at one instant of time?

   If so, in which frame number (or numbers)? _____

   Draw a vertical line through your graphs of part a to indicate this instant of time.

7. Below are four position-versus-time graphs. For each, draw the corresponding velocity-versus-time graph directly below it. A vertical line drawn through both graphs should connect the velocity v_x at time t with the position x at the *same* time t. There are no numbers, but your graphs should correctly indicate the *relative* speeds.

   a. 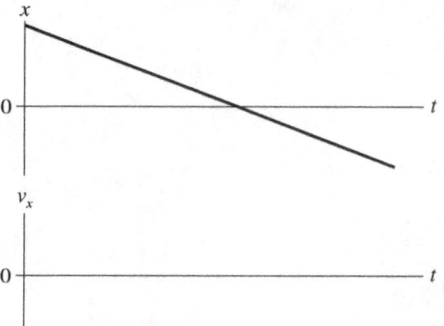   b.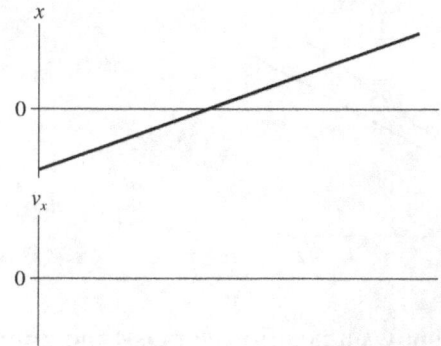

   c. 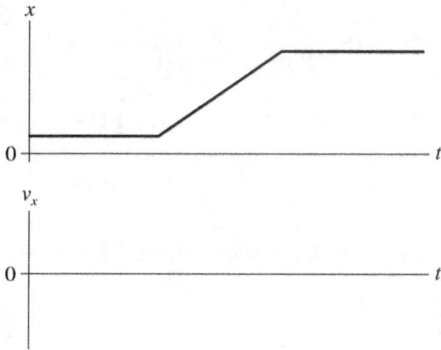   d.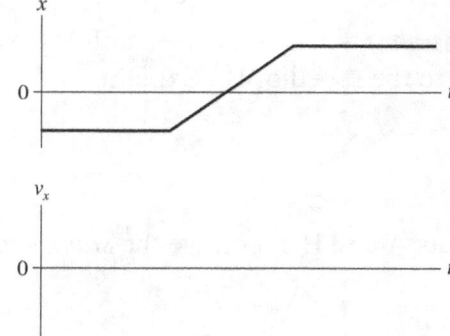

8. Below are two velocity-versus-time graphs. For each:
   - Draw the corresponding position-versus-time graph.
   - Give a written description of the motion.

   Assume that the motion takes place along a horizontal line and that $x_i = 0$.

   a.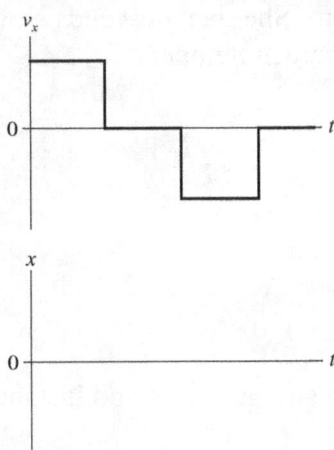

   b.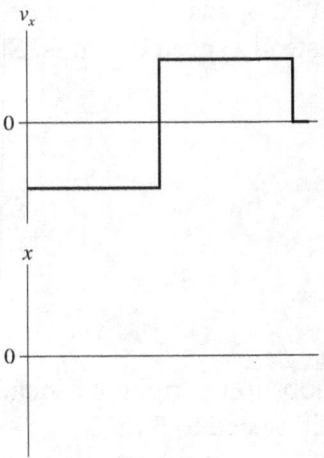

9. The figure shows a position-versus-time graph for a moving object. At which lettered point or points

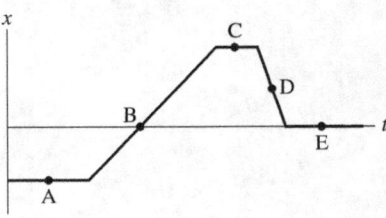

   a. Is the object moving the slowest without being at rest? _____

   b. Is the object moving the fastest? _____

   c. Is the object at rest? _____

   d. Does the object have a constant nonzero velocity? _____

   e. Is the object moving to the left? _____

2.2 Uniform Motion

10. Sketch position-versus-time graphs for the following motions. Include appropriate numerical scales along both axes. A small amount of computation may be necessary.

   a. A parachutist opens her parachute at an altitude of 1500 m. She then descends slowly to earth at a steady speed of 5 m/s. Start your graph as her parachute opens.

   b. A rabbit hops to the right at a steady 1 m/s for 4 s. It then sees a coyote and instantly increases its speed to 3 m/s.

   c. Quarterback Bill throws the ball to the right at a speed of 15 m/s. It is intercepted 45 m away by Carlos, who is running to the left at 7.5 m/s. Carlos carries the ball 60 m to score. Let $x = 0$ m be the point where Bill throws the ball. Draw the graph for the *football*.

11. The height of a building is proportional to the number of stories it has. If a 25-story building is 260 ft tall, what is the height of an 80-story building? Answer this using ratios, not by calculating the height per story.

2.3 Instantaneous Velocity

12. Below are two position-versus-time graphs. For each, draw the corresponding velocity-versus-time graph directly below it. A vertical line drawn through both graphs should connect the velocity v_x at time t with the position x at the *same* time t. There are no numbers, but your graphs should correctly indicate the *relative* speeds.

a.
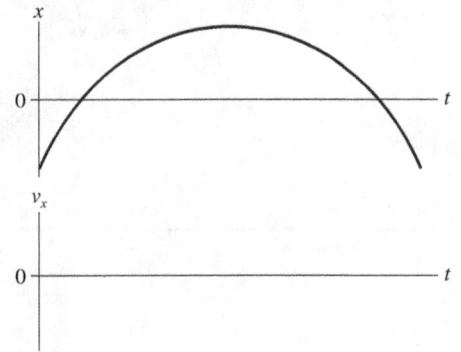

b.
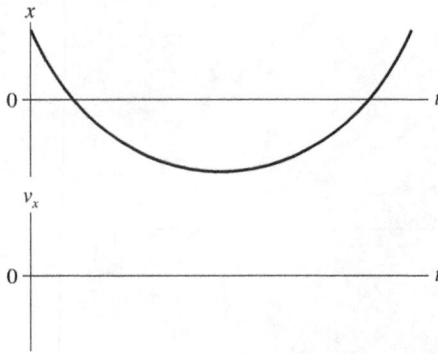

13. The figure shows six frames from the motion diagram of two moving cars, A and B.

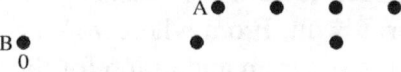

   a. Draw both a position-versus-time graph and a velocity-versus-time graph. Show *both* cars on each graph. Label them A and B.

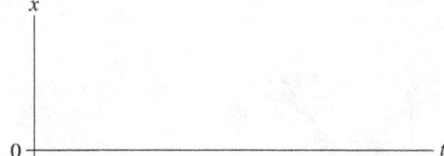

   b. Do the two cars ever have the same position at one instant of time? _____

   If so, in which frame number (or numbers)? _____

   Draw a vertical line through your graphs of part a to indicate this instant of time.

   c. Do the two cars ever have the same velocity at one instant of time? _____

   If so, between which two frames? _____

14. For each of the following motions, draw
    - A motion diagram and
    - Both position and velocity graphs.

    a. A car starts from rest, steadily speeds up to 40 mph in 15 s, moves at a constant speed for 30 s, and then comes to a halt in 5 s.

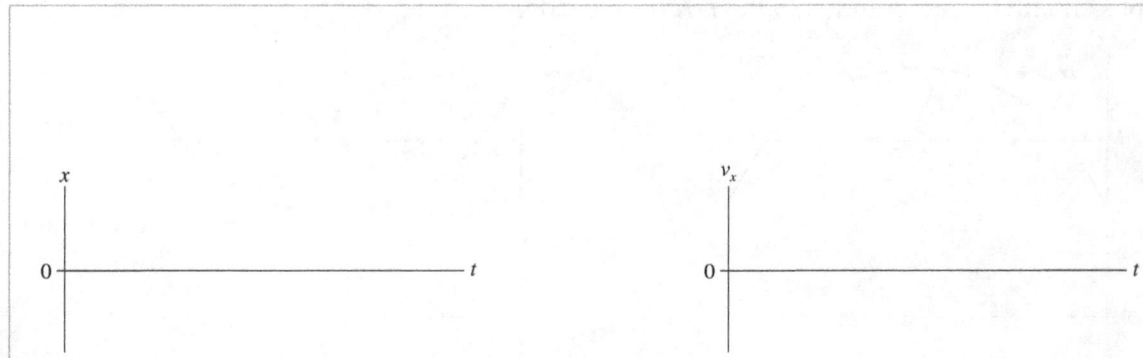

    b. A pitcher throws a baseball with a speed of 40 m/s. One-half second later, the batter hits the ball back toward the pitcher, just missing him, with a speed of 60 m/s. The ball is caught 1 s after it is hit. From where you are sitting, the batter is to the right of the pitcher. Draw your motion diagram and graph for the *horizontal* motion of the ball.

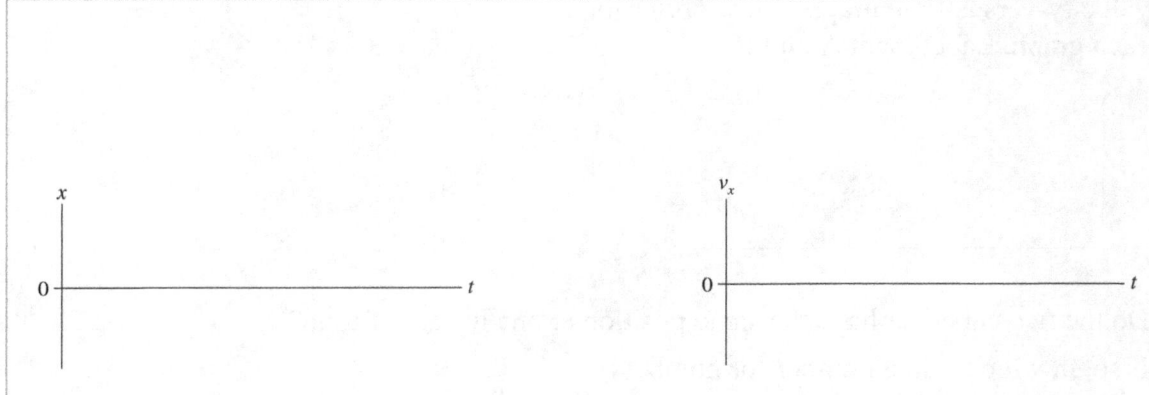

15. Which of the position-versus-time graphs below goes with the velocity-versus-time graph shown at the right? Circle the part label of your choice.

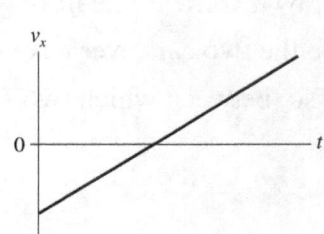

a. 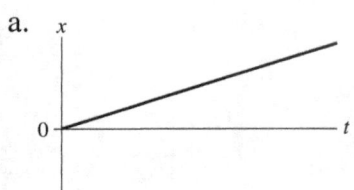 b. 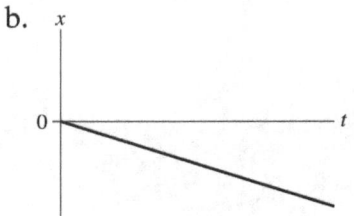 c.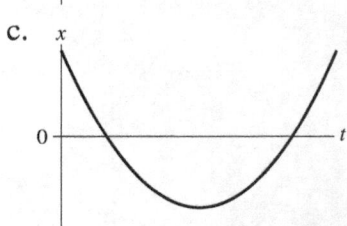

2.4 Acceleration

16. The four motion diagrams below show an initial point 0 and a final point 1. A pictorial representation would define the five symbols: x_0, x_1, v_{0x}, v_{1x}, and a_x for horizontal motion and equivalent symbols with y for vertical motion. Determine whether each of these quantities is positive, negative, or zero. Give your answer by writing $+$, $-$, or 0 in the table below.

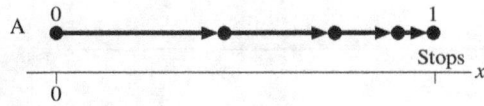
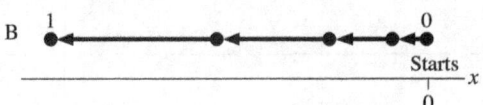

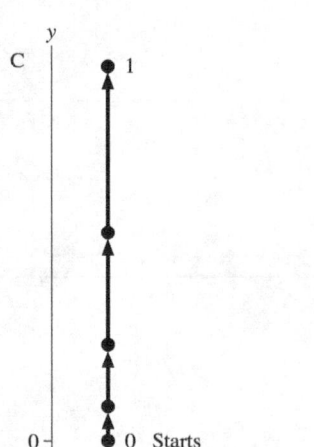
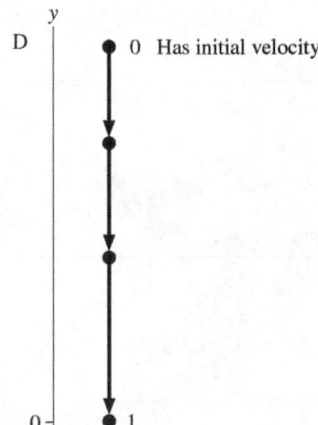

	A	B	C	D
x_0 or y_0				
x_1 or y_1				
v_{0x} or v_{0y}				
v_{1x} or v_{1x}				
a_x or a_y				

17. The three symbols x, v_x, and a_x have eight possible combinations of *signs*. For example, one combination is $(x, v_x, a_x) = (+, -, +)$.

    a. List all eight combinations of signs for x, v_x, a_x.

    1. _____   5. _____

    2. _____   6. _____

    3. _____   7. _____

    4. _____   8. _____

b. For each of the eight combinations of signs you identified in part a:
- Draw a four-dot motion diagram of an object that has these signs for x, v_x, and a_x.
- Draw the diagram *above* the axis whose number corresponds to part a.
- Use **black** and **red** for your $\vec{v}$ and $\vec{a}$ vectors. Be sure to label the vectors.

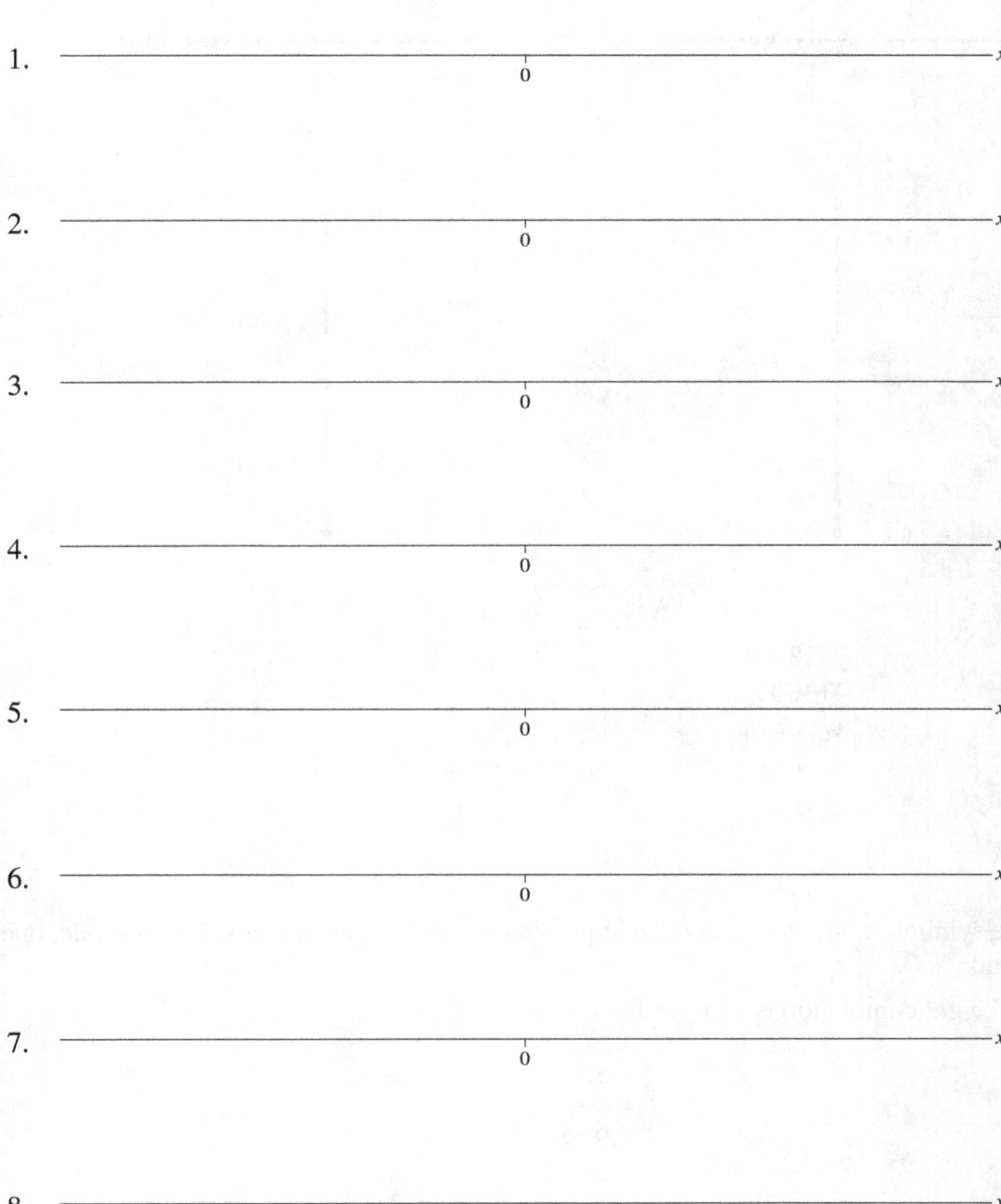

2.5 Motion with Constant Acceleration

18. Draw a motion diagram to illustrate each of the following situations.

    a. $a_x = 0$ but $v_x \neq 0$.

    b. $v_x = 0$ but $a_x \neq 0$.

    c. $v_x < 0$ and $a_x > 0$.

19. The quantity y is proportional to the square of x, and $y = 36$ when $x = 3$.

    a. Find y if $x = 5$.   b. Find x if $y = 16$.

    c. By what *factor* must x increase for the value of y to double?

    d. Consider the equation in your text relating Δx and Δt for motion with constant acceleration a_x. Which of these three quantities plays the role of x in a quadratic relationship $y = Ax^2$? Which plays the role of y?

20. Below are three velocity-versus-time graphs. For each, draw the corresponding acceleration-versus-time graph.

    a. 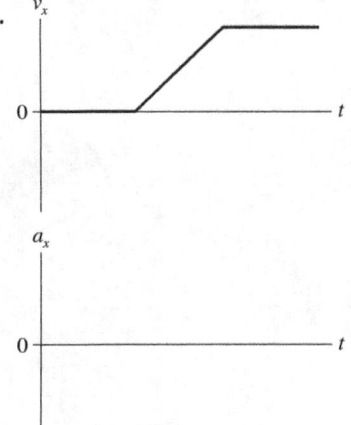   b. 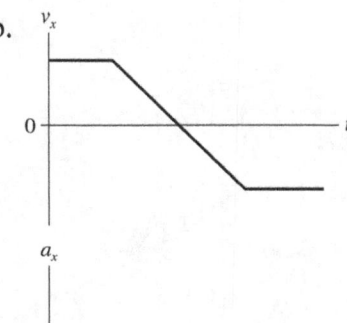   c.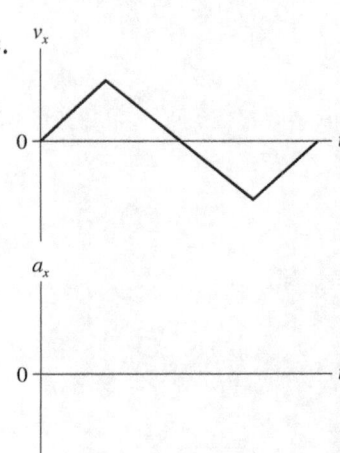

2.6 Solving One-Dimensional Motion Problems

21. Draw a pictorial representation of each situation described below. That is, (i) sketch the situation, showing appropriate points in the motion, (ii) establish a coordinate system on your sketch, and (iii) define appropriate symbols for the known and unknown quantities. **Do not solve.** See textbook Example 2.10 as an example.

    a. A car traveling at 30 m/s screeches to a halt, leaving 55-m-long skid marks. What was the car's acceleration while braking?

    b. A bicyclist starts from rest and accelerates at 4.0 m/s^2 for 3.0 s. The cyclist then travels for 20 s at a constant speed. How far does the cyclist travel?

    c. You are driving your car at 12 m/s when a deer jumps in front of your car. What is the shortest stopping distance for your car if your reaction time is 0.80 s and your car brakes at 6.0 m/s^2?

2.7 Free Fall

22. A ball is thrown straight up into the air. At each of the following instants, is the ball's vertical acceleration described by $a_y = -g$, $-g < a_y < 0$, $a_y = 0$, $0 < a_y < g$, $a_y = g$, or $a_y > g$?

    a. Just after leaving your hand? _____

    b. At the very top (maximum height)? _____

    c. Just before hitting the ground? _____

23. A ball is thrown straight up into the air from $y = 0$. It reaches height h, and then falls back down to the person's hand. On the axes below, graph the ball's position, velocity, and acceleration from an instant after it leaves the thrower's hand until an instant before it is caught. Make sure the information on your three graphs is vertically aligned.

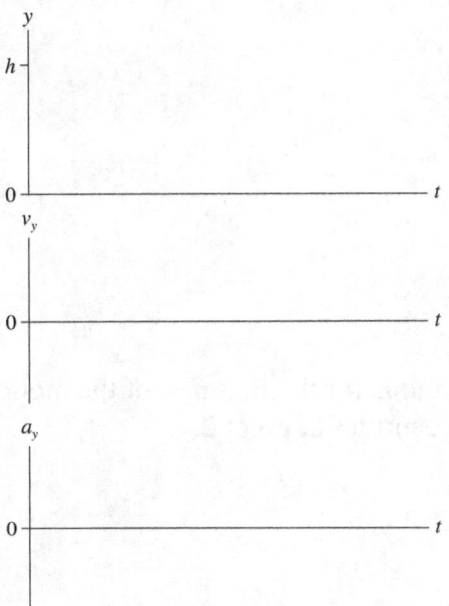

24. A rock is *thrown* (not dropped) straight down from a bridge into the river below.

    a. Immediately *after* being released, is the magnitude of the rock's acceleration greater than g, less than g, or equal to g? Explain.

    b. Immediately before hitting the water, is the magnitude of the rock's acceleration greater than g, less than g, or equal to g? Explain.

25. A model rocket is launched straight up with constant acceleration a. It runs out of fuel at time t.
PSA Suppose you need to determine the maximum height reached by the rocket. We'll assume that
2.1 air resistance is negligible.

   a. Is the rocket at maximum height the instant it runs out of fuel? _____

   b. Is there anything other than gravity acting on the rocket after it runs out of fuel? _____

   c. What is the name of motion under the influence of only gravity? _____

   d. Draw a pictorial representation for this problem. You should have three identified points in the motion: launch, out of fuel, maximum height. Call these points 1, 2, and 3.

     • Using subscripts, define 11 quantities: y, v_y, and t at each of the three points, plus acceleration a_1 connecting points 1 and 2 and acceleration a_2 connecting points 2 and 3.

     • 7 of these quantities are Knowns; identify them and specify their values. Some are 0. Others can be given in terms of a, t, and g, which are "known," because they're stated in the problem, even though you don't have numerical values for them. For example, $t_1 = t$. Be careful with signs!

     • Identify which one of the 4 unknown quantities you're trying to find.

   e. This is a two-part problem. Write two kinematic equations for the first part of the motion to determine—again symbolically—the two unknown quantities at point 2.

   f. Now write a kinematic equation for the second half of the motion that will allow you to find the desired unknown that will answer the question. Just write the equation; don't yet solve it.

   g. Now, substitute what you learned in part e into your equation of part f, do the algebra to solve for the unknown, and simplify the result as much as possible.

You Write the Problem!

Exercises 26–28: You are given the kinematic equation that is used to solve a problem. For each of these:

   a. Write a *realistic* physics problem for which this is the correct equation. Look at worked examples and end-of-chapter problems in the textbook to see what realistic physics problems are like. Be sure that the problem you write, and the answer you ask for, is consistent with the information given in the equation.

   b. Draw a pictorial representation for your problem.

   c. Finish the solution of the problem.

26. $64 \text{ m} = 0.0 \text{ m} + (32 \text{ m/s})(4.0 \text{ s} - 0.0 \text{ s}) + \frac{1}{2}a_x(4.0 \text{ s} - 0.0 \text{ s})^2$

27. $0.0 \text{ m/s} = 36 \text{ m/s} - (3.0 \text{ m/s}^2)t$

28. $(10 \text{ m/s})^2 = (v_y)_i^2 - 2(9.8 \text{ m/s}^2)(10 \text{ m} - 0 \text{ m})$

3 Vectors and Motion in Two Dimensions

3.1 Using Vectors

Exercises 1–3: Draw and label the vector sum $\vec{A} + \vec{B}$ or $\vec{A} + \vec{B} + \vec{C}$.

1.
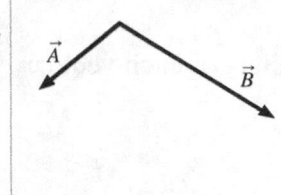

2.
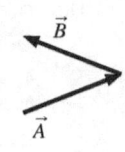

3.
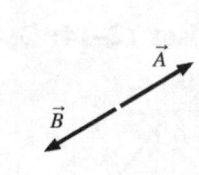

4. Draw and label the vector $2\vec{A}$ and the vector $\frac{1}{2}\vec{A}$.

Exercises 5–7: Draw and label the vector difference $\vec{A} - \vec{B}$.

5.
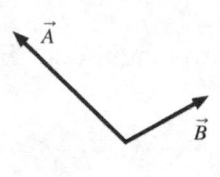

6.
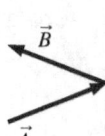

7.
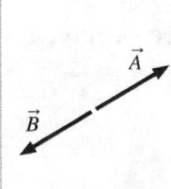

8. Given vectors $\vec{A}$ and $\vec{B}$ below, find the vector $\vec{C} = 2\vec{A} - 3\vec{B}$.

 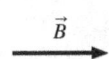

3.2 Coordinate Systems and Vector Components

Exercises 9–11: Draw and label the x- and y-component vectors of the vector shown.

9.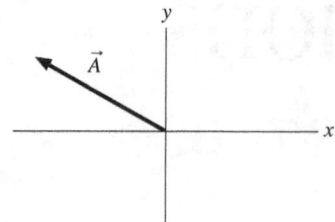

10.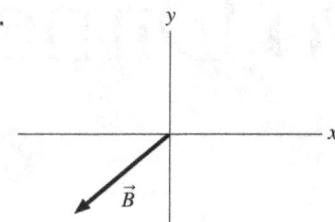

11.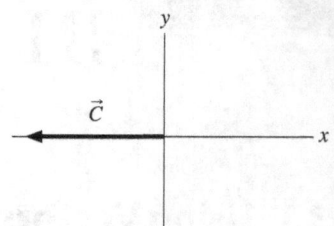

Exercises 12–14: Determine the numerical values of the x- and y-components of each vector.

12.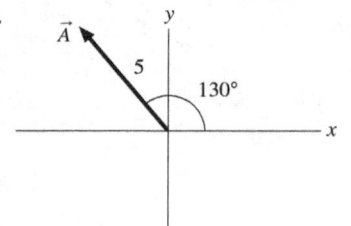

$A_x =$ _____

$A_y =$ _____

13.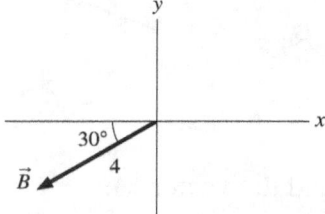

$B_x =$ _____

$B_y =$ _____

14.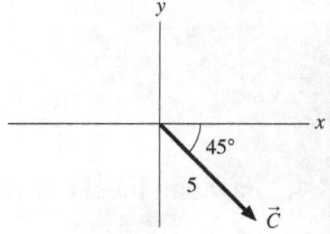

$C_x =$ _____

$C_y =$ _____

15. What is the vector sum $\vec{D} = \vec{A} + \vec{B} + \vec{C}$ of the three vectors defined in Exercises 12–14?

$D_x =$ _____ $D_y =$ _____

16. Can a vector have a component equal to zero and still have nonzero magnitude? Explain.

17. Can a vector have zero magnitude if one of its components is nonzero? Explain.

Exercises 18–20: For each vector:
- Draw the vector on the axes provided.
- Draw and label an angle θ of magnitude $\leq 90°$ to describe the direction of the vector.
- Find the magnitude and the angle of the vector.

18. $A_x = 3$, $A_y = -2$

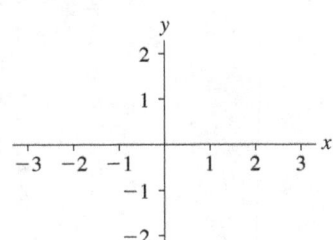

$A =$ _____

$\theta =$ _____

19. $B_x = -2$, $B_y = 2$

$B =$ _____

$\theta =$ _____

20. $C_x = 0$, $C_y = -2$

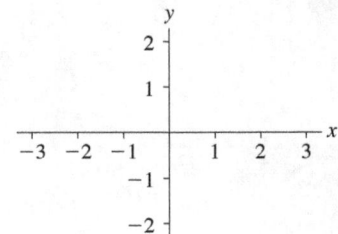

$C =$ _____

$\theta =$ _____

Exercises 21–23: Define vector $\vec{A} = (5, 30°$ above the horizontal$)$. Determine the components A_x and A_y in the three coordinate systems shown below. Show your work below the figure.

21.

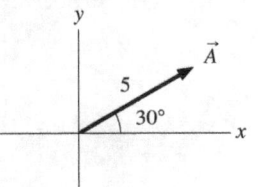

$A_x =$ _____

$A_y =$ _____

22.

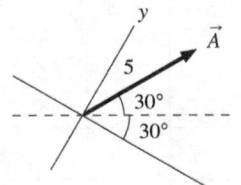

$A_x =$ _____

$A_y =$ _____

23.

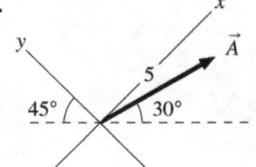

$A_x =$ _____

$A_y =$ _____

3.3 Motion on a Ramp

24. Draw vector arrows on the figure at each of the five dots to show the ball's acceleration at that point, or write $\vec{a} = \vec{0}$ if appropriate. The length of each arrow should be proportional to the magnitude of the acceleration at that point.

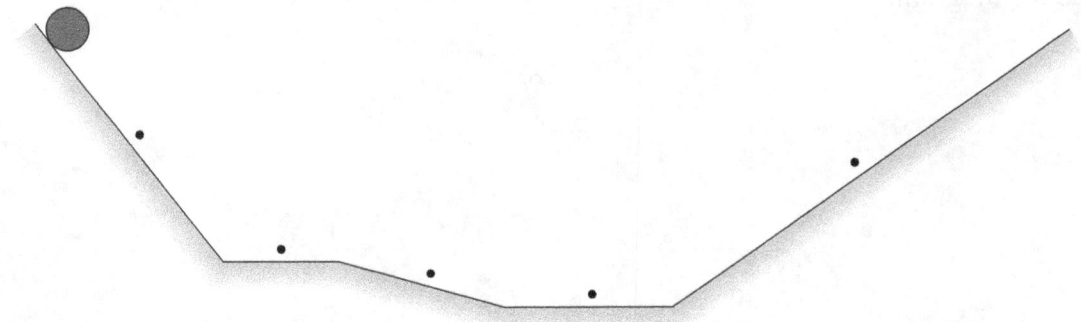

3.4 Motion in Two Dimensions

25. The figure below shows the positions of a moving object at three successive points in a motion diagram. Draw and label the velocity vector $\vec{v}_0$ for the motion from 0 to 1 and the vector $\vec{v}_1$ for the motion from 1 to 2. Then determine and draw the vector $\vec{v}_1 - \vec{v}_0$ with its tail on point 1.

26. A car enters an icy intersection traveling 16 m/s due north. After a collision with a truck, the car slides away moving 12 m/s due east. Draw arrows on the picture below to show (i) the car's velocity $\vec{v}_0$ when entering the intersection, (ii) its velocity $\vec{v}_1$ when leaving, and (iii) the car's change in velocity $\Delta\vec{v} = \vec{v}_1 - \vec{v}_0$ due to the collision.

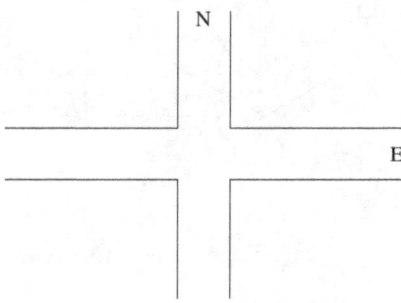

Exercises 27–28: The figures below show an object's position at three successive points in a motion diagram. For each diagram:
- Draw and label the initial and final velocity vectors $\vec{v}_0$ and $\vec{v}_1$. Use **black**.
- Use the steps of Tactics Box 3.3 to find the change in velocity $\Delta\vec{v}$.
- Draw and label $\vec{a}$ at the proper location on the motion diagram. Use **red**.
- Is the object speeding up, slowing down, or moving at a constant speed? Write your answer beside the diagram.

27.   28.

3.5 Projectile Motion

29. Complete the motion diagram for this trajectory, showing velocity and acceleration vectors.

30. A projectile is launched over level ground and lands some distance away.

    a. Is there any point on the trajectory where $\vec{v}$ and $\vec{a}$ are parallel to each other? If so, where?

    b. Is there any point where $\vec{v}$ and $\vec{a}$ are perpendicular to each other? If so, where?

    c. Which of the following remain constant throughout the flight: x, y, v, v_x, v_y, a_x, a_y?

31. A ball is projected horizontally at 10 m/s and hits the ground 2.0 s later. The figure shows the ball's position every 0.5 s.
    - At each dot, starting with $t = 0.5$ s, draw a vector for the horizontal component of velocity v_x and a vector for the vertical component of velocity v_y.
    - The length of each vector should indicate its magnitude, using the length of the 10 m/s vector at $t = 0$ s as a reference.

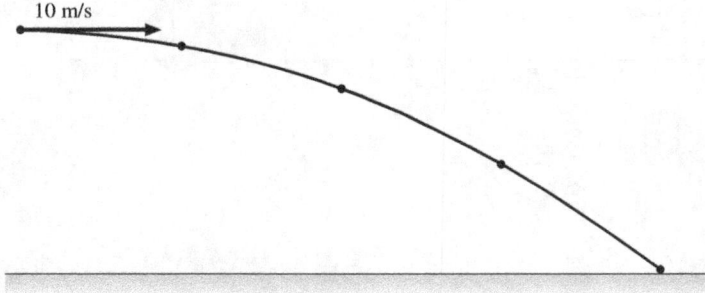

3.6 Projectile Motion: Solving Problems

32. a. A cart rolling at constant velocity fires a ball straight up. When the ball comes down, will it land in front of the launching tube, behind the launching tube, or directly in it? Explain.

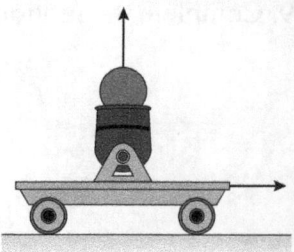

   b. Will your answer change if the cart is accelerating in the forward direction? If so, how?

33. Figure 3.32 in the textbook shows that a projectile launched at a 60° angle has the same range as a projectile launched with the same speed at a 30° angle. If these two projectiles are launched simultaneously, do they hit the ground simultaneously? If not, which lands first? Explain.

34. Rank in order, from shortest to longest, the amount of time it takes each of these projectiles to hit the ground. (Some may be simultaneous.)

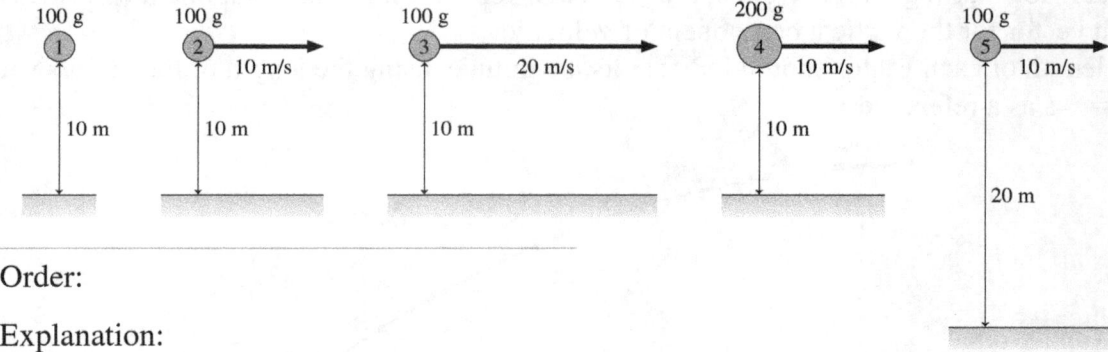

   Order:

   Explanation:

3.7 Circular Motion

35. The dots of a motion diagram are shown below for an object in uniform circular motion. Carefully complete the diagram.
    - Draw and label the velocity vectors $\vec{v}$. Use a **black** pen or pencil.
    - Draw and label the acceleration vectors $\vec{a}$. Use a **red** pen or pencil.

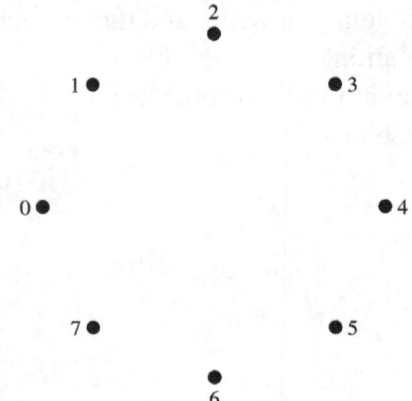

36. The figure is a bird's-eye view of a car traveling at a steady 20 mph. At each of the three dots, labeled A, B, and C, either (a) write $\vec{a} = \vec{0}$ if the car is not accelerating or (b) draw and label a vector, with its tail at the dot, to show the car's acceleration.

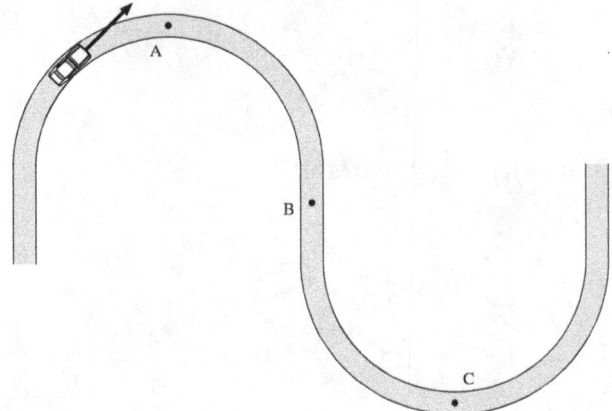

37. An object travels in a circle of radius r at constant speed v.

    a. By what factor does the object's acceleration change if its speed is doubled and the radius is unchanged?

    b. By what factor does the acceleration change if the radius of the circle is doubled and its speed is unchanged?

You Write the Problem!

Exercises 38–40: You are given the equation or equations used to solve a problem. For each of these:

a. Write a realistic physics problem for which this is the correct equation. Look at worked examples and end-of-chapter problems in the textbook to see what realistic physics problems are like. Be sure that the problem you write, and the answer you ask for, is consistent with the information given in the equation.

b. Draw the pictorial representation for your problem.

c. Finish the solution of the problem.

38. $x_1 = 0 \text{ m} + (30 \text{ m/s})t_1$

$0 \text{ m} = 300 \text{ m} - \frac{1}{2}(9.8 \text{ m/s}^2)t_1^2$

39. $x_1 = 0 \text{ m} + (50 \cos 30° \text{ m/s})t_1$

$0 \text{ m} = 0 \text{ m} + (50 \sin 30° \text{ m/s})t_1 - \frac{1}{2}(9.8 \text{ m/s}^2)t_1^2$

40. $2.5 \text{ m/s}^2 = \dfrac{v^2}{10 \text{ m}}$

3.8 Relative Motion

41. On a ferry moving steadily forward in still water at 5 m/s, a passenger walks toward the back of the boat at 2 m/s. (i) Write a symbolic equation to find the velocity of the passenger with respect to the water using $(v_x)_{AB}$ notation. (ii) Substitute the appropriate values into your equation to determine the value of that velocity.

42. A boat crossing a river can move at 5 m/s with respect to the water. The river is flowing to the right at 3 m/s. In (a), the boat points straight across the river and is carried downstream by the water. In (b), the boat is angled upstream by the amount needed for it to travel straight across the river. For each situation, draw the velocity vectors $\vec{v}_{RS}$ of the river with respect to the shore, $\vec{v}_{BR}$ of the boat with respect to the river, and $\vec{v}_{BS}$ of the boat with respect to the shore.

a.
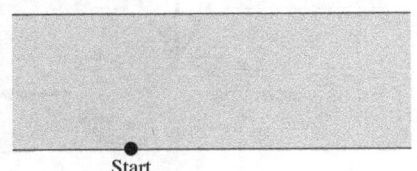
Start

b.
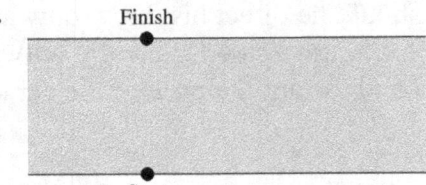
Finish

Start

43. Ryan, Samantha, and Tomas are driving their convertibles. At the same instant, they each see a jet plane with an instantaneous velocity of 200 m/s and an acceleration of 5 m/s². Rank in order, from largest to smallest, the jet's *speed* v_R, v_S, and v_T according to Ryan, Samantha, and Tomas. Explain.

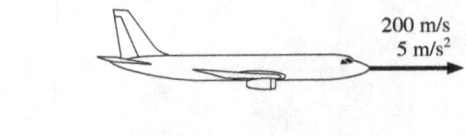
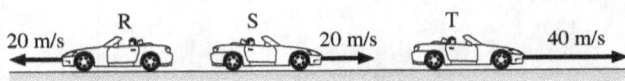

Order:

Explanation:

44. An electromagnet on the ceiling of an airplane holds a steel ball. When a button is pushed, the magnet releases the ball. The experiment is first done while the plane is parked on the ground, and the point where the ball hits the floor is marked with an X. Then the experiment is repeated while the plane is flying level at a steady 500 mph. Does the ball land slightly in front of the X (toward the nose of the plane), on the X, or slightly behind the X (toward the tail of the plane)? Explain.

45. Zack is driving past his house. He wants to toss his physics book out the window and have it land in his driveway. If he lets go of the book exactly as he passes the end of the driveway, should he direct his throw outward and toward the front of the car (throw 1), straight outward (throw 2), or outward and toward the back of the car (throw 3)? Explain.

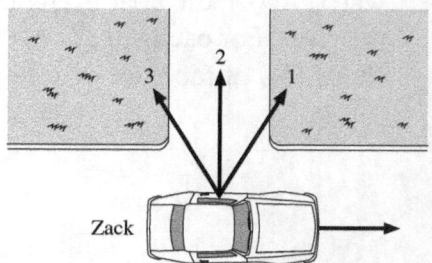

46. Yvette and Zack are driving down the freeway side by side with their windows rolled down. Zack wants to toss his physics book out the window and have it land in Yvette's front seat. Should he direct his throw outward and toward the front of the car (throw 1), straight outward (throw 2), or outward and toward the back of the car (throw 3)? Explain.

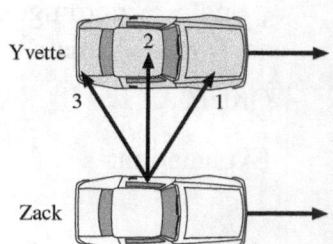

4 Forces and Newton's Laws of Motion

4.1 Motion and Force

1. Using the particle model, draw the vector of the force a person exerts on a table when (a) pulling it to the right across a level floor with a force of magnitude F, (b) pulling it to the left across a level floor with force $2F$, and (c) *pushing* it to the right across a level floor with force F.

a. Table pulled right
   with force F

b. Table pulled left
   with force $2F$

c. Table pushed right
   with force F

2. Two or more forces are shown on the objects below. Draw and label the net force $\vec{F}_{net}$.

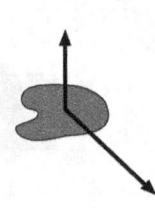

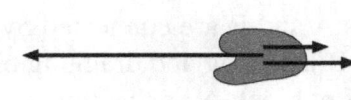

3. Two or more forces are shown on the objects below. Draw and label the net force $\vec{F}_{net}$.

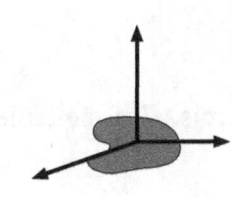

4.2 A Short Catalog of Forces

4.3 Identifying Forces

Exercises 4–8: Follow the six-step procedure of Tactics Box 4.2 to identify and name all the forces acting on the object.

4. An elevator suspended by a cable is descending at constant velocity.

5. A compressed spring is pushing a block across a rough horizontal table.

6. A brick is falling from the roof of a three-story building.

7. Blocks A and B are connected by a string passing over a pulley. Block B is falling and dragging block A across a frictionless table. Let block A be the object for analysis.

8. Repeat Exercise 7 if the table has friction.

4.4 What Do Forces Do?

9. The figure shows an acceleration-versus-force graph for an object of mass m. Data has been plotted as individual points, and a line has been drawn through the points.

   Draw and label, directly on the figure, the acceleration-versus-force graphs for objects of mass

   a. $2m$ b. $0.5m$

   Use triangles ▲ to show four points for the object of mass $2m$, then draw a line through the points. Use squares ■ for the object of mass $0.5m$.

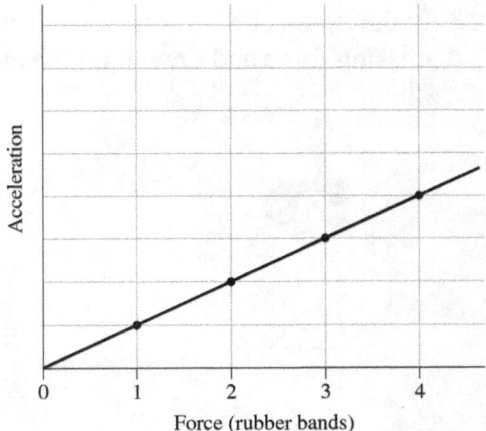

10. The quantity y is inversely proportional to x, and $y = 4$ when $x = 9$.

    a. Write an equation to represent this inverse relationship for all y and x.

    b. Find y if $x = 12$. _____   c. Find x if $y = 36$. _____

    d. Compare your equation in part a to Newton's second law, $a = \dfrac{F}{m}$.

        Which quantity assumes the role of x? _____

        Which quantity assumes the role of y? _____

        What is the constant of proportionality relating a and m? _____

11. The quantity y is inversely proportional to x. For one value of x, $y = 12$.

    a. What is the value of y if x is doubled? _____

    b. What is the value of y if the original value of x is halved? _____

12. A steady force applied to a 2.0 kg mass causes it to accelerate at 4.0 m/s². The same force is then applied to a 4.0 kg mass. Use ratio reasoning to find the acceleration of the 4.0 kg mass.

4.5 Newton's Second Law

13. In the figures below, one force is missing. Use the given direction of acceleration to determine the missing force and draw it on the object. Do all work directly on the figure.

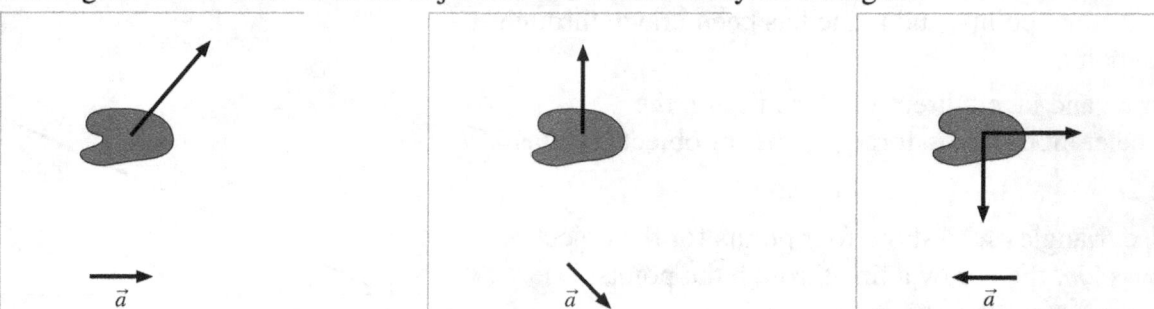

14. Forces are shown on three objects. For each:
    a. Draw and label the net force vector. Do this right on the figure.
    b. Below the figure, draw and label the object's acceleration vector, as was done in Exercise 13.

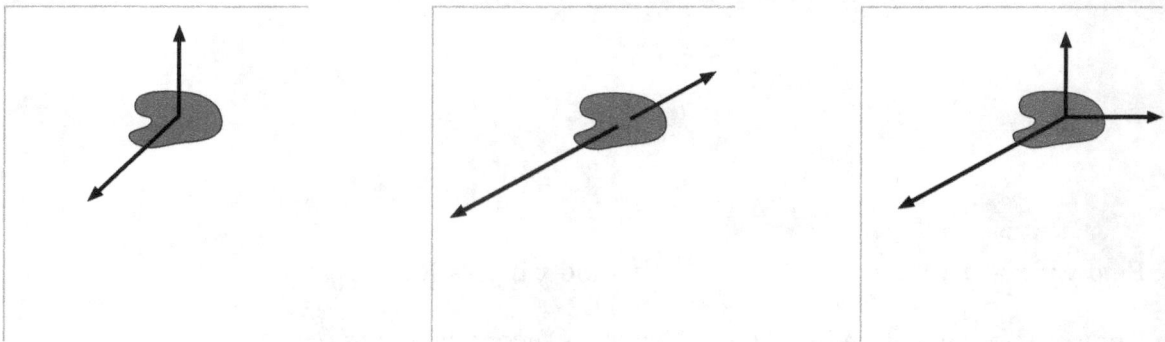

15. Below are two motion diagrams for a particle. Draw and label the net force vector at point 2.

16. A constant force applied to an object causes the object to accelerate at 10 m/s^2. What will the acceleration of this object be if
    a. The force is doubled? _____   b. The mass is doubled? _____

    c. The force is doubled *and* the mass is doubled? _____

    d. The force is doubled *and* the mass is halved? _____

4.6 Free-Body Diagrams

Exercises 17–20:
- Draw a picture and identify the forces, following Tactics Box 4.2, then
- Draw a free-body diagram for the object, following each of the steps given in Tactics Box 4.3. Be sure to think carefully about the direction of $\vec{F}_{net}$.

Note: Draw individual force vectors with a **black** or **blue** pencil or pen. Draw the *net* force vector $\vec{F}_{net}$ with a **red** pencil or pen.

17. A heavy crate is being lowered straight down at a constant speed by a steel cable.

18. A boy is pushing a box across a rough floor at a steadily increasing speed. Let the box be the object for analysis.

19. A bicycle is speeding up down a hill. Friction is negligible, but air resistance is not.

20. You've slammed on your car brakes while going down a hill. The car is skidding to a halt.

Exercises 21–22: Identify the forces and draw a free-body diagram for the object of interest.

21. You are going to toss a rock *straight up* into the air by placing it on the palm of your hand (you're not gripping it), then pushing your hand up very rapidly. You may want to toss an object into the air this way to help you think about the situation. The rock is the object of interest.

    a. As you hold the rock at rest on your palm, before moving your hand.

    b. As your hand is moving up but before the rock leaves your hand.

    c. One-tenth of a second after the rock leaves your hand.

    d. After the rock has reached its highest point and is now falling straight down.

22. Block B has just been released and is beginning to fall. Block A is the object of interest. The table has friction.

4.7 Newton's Third Law

Exercises 23–25: Each of the following situations has two or more interacting objects. Draw a picture similar to Figure 4.30 in the textbook in which you
- Show the interacting objects, with a small gap separating them.
- Draw the force vectors of all action/reaction pairs.
- Label the force vectors, using a notation like $\vec{F}_{A \text{ on } B}$ and $\vec{F}_{B \text{ on } A}$.

23. A bat hits a ball. Draw your picture from the perspective of someone seeing the *end* of the bat at the moment it strikes the ball. The objects are the bat and the ball.

24. A girl pushes a box across the floor. Friction is not negligible. The objects are the girl, the box, and the floor.

25. A crate is in the back of a truck as the truck accelerates forward. The crate does not slip. The objects are the truck, the crate, and the ground.

26. You find yourself in the middle of a frozen lake with a surface so slippery that you cannot walk. However, you happen to have several rocks in your pocket. The ice is extremely hard. It cannot be chipped, and the rocks slip on it just as much as your feet do. Can you think of a way to get to shore? Use pictures, forces, and Newton's laws to explain your reasoning.

27. How do basketball players jump straight up into the air? Your explanation should include pictures showing forces on the player and forces on the ground.

You Write the Problem!

Exercises 28–32: You are given the free-body diagram of an object with one or more forces acting on it. For each of these:

    a. Determine the direction of the object's acceleration $\vec{a}$. Draw and label the acceleration vector next to the free-body diagram. Or, if appropriate, write $\vec{a} = \vec{0}$

    b. Write a short description of a real object for which this is the correct free-body diagram. Use the worked examples in the textbook as models of what a description should be like.

28.

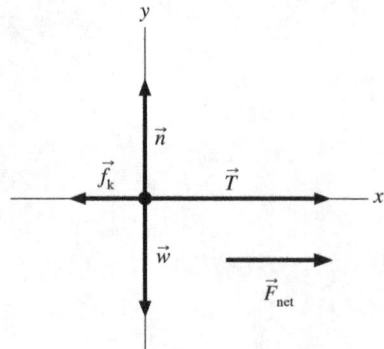

29.

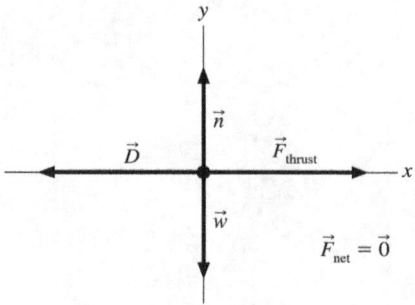

30.

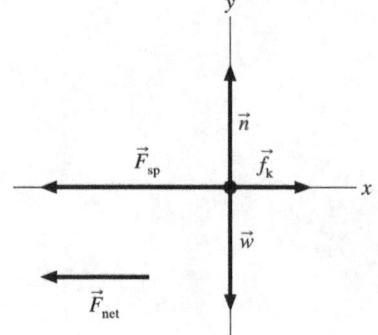

31.

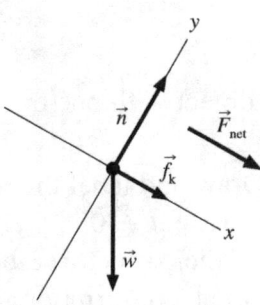

32.

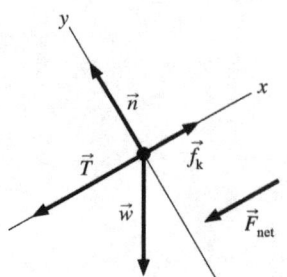

5 Applying Newton's Laws

5.1 Equilibrium

1. If an object is at rest, can you conclude that there are no forces acting on it? Explain.

2. If a force is exerted on an object, is it possible for that object to be moving with constant velocity? Explain.

3. A hollow tube forms three-quarters of a circle. It is lying flat on a table. A ball is shot through the tube at high speed. As the ball emerges from the other end, does it follow path A, path B, or path C? Explain your reasoning.

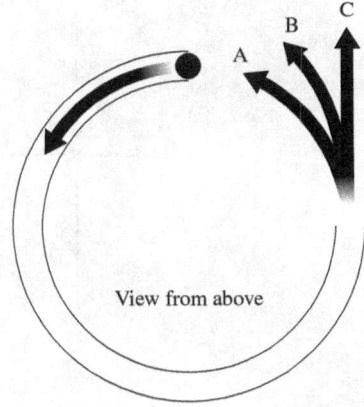

View from above

4. The vectors below show five forces that can be applied individually or in combinations to an object. Which forces or combinations of forces will cause the object to be in equilibrium?

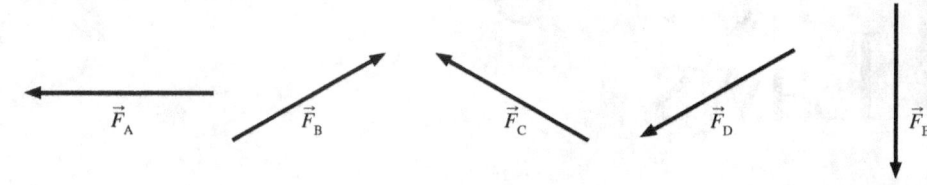

5. The free-body diagrams show a force or forces acting on an object. Draw and label one more force (one that is appropriate to the situation) that will cause the object to be in equilibrium.

a.

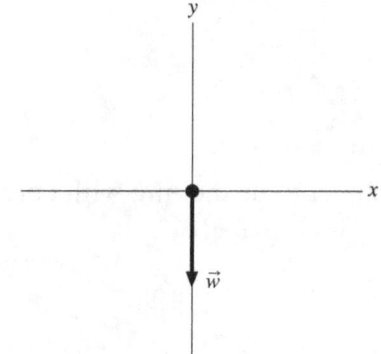

b.

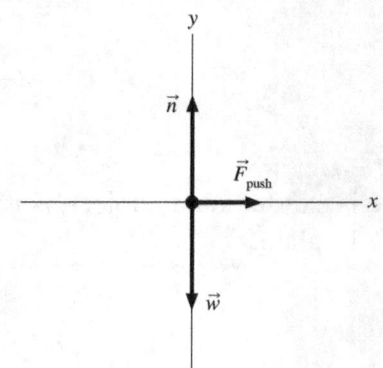

6. The free-body diagrams show a force or forces acting on an object. Draw and label one more force (one that is appropriate to the situation) that will cause the object to be in equilibrium.

a.

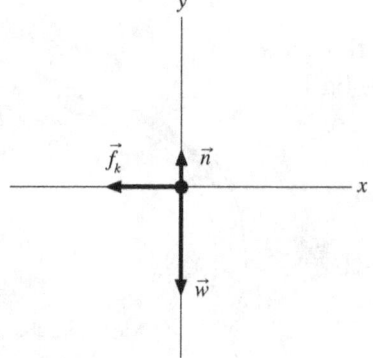

b.

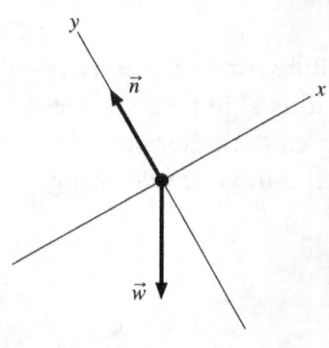

5.2 Dynamics and Newton's Second Law

7. a. An elevator travels *upward* at a constant speed. The elevator hangs by a single cable. Friction and air resistance are negligible. Is the tension in the cable greater than, less than, or equal to the weight of the elevator? Explain. Your explanation should include both a free-body diagram and reference to appropriate laws of physics.

b. The elevator travels *downward* and is slowing down. Is the tension in the cable greater than, less than, or equal to the weight of the elevator? Explain.

Exercises 8–9: The figures show free-body diagrams for an object of mass m. Write the x- and y-components of Newton's second law. Write your equations in terms of the *magnitudes* of the forces $F_1, F_2, \ldots$ and any *angles* defined in the diagram. One equation is shown in each question to illustrate the procedure.

8.

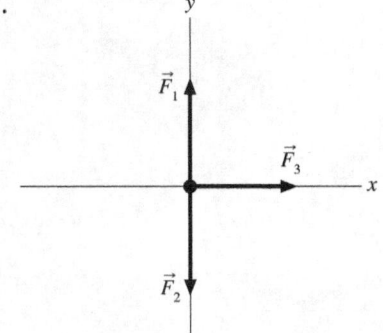

$ma_x =$

$ma_y = F_1 - F_2$

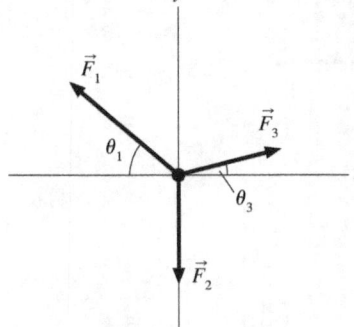

$ma_x =$

$ma_y =$

9.

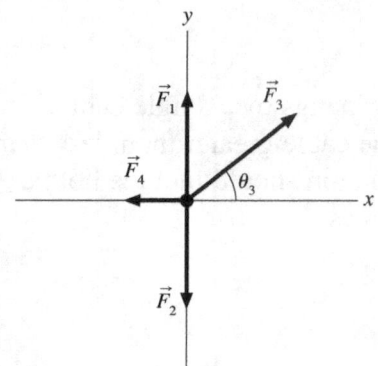
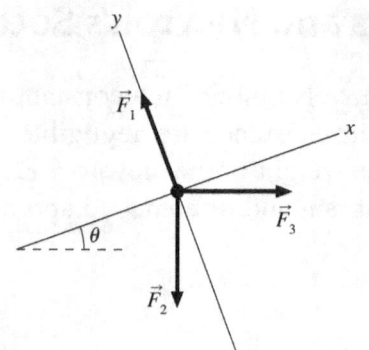

$$ma_x = F_3 \cos \theta_3 - F_4$$

$$ma_y =$$

$$ma_x =$$

$$ma_y =$$

Exercises 10–12: Two or more forces, shown on a free-body diagram, are exerted on a 2 kg object. The units of the grid are newtons. For each:
- Draw a vector arrow *on the grid*, starting at the origin, to show the net force $\vec{F}_{net}$.
- In the space to the right, determine the numerical values of the components a_x and a_y.

10.

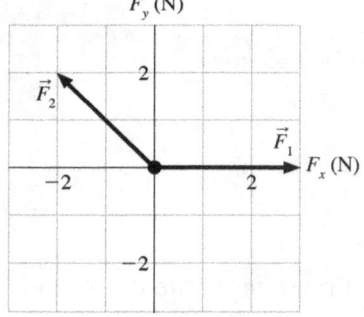

$$a_x =$$

$$a_y =$$

11.

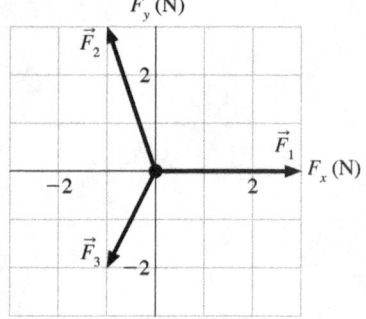

$$a_x =$$

$$a_y =$$

12.

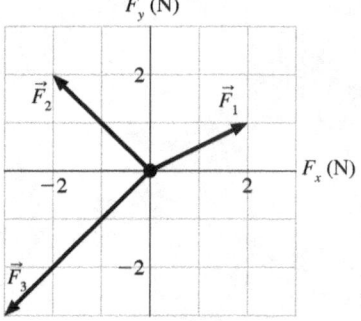

$$a_x =$$

$$a_y =$$

Exercises 13–15: Three forces $\vec{F}_1$, $\vec{F}_2$, and $\vec{F}_3$ acting together cause a 1 kg object to accelerate with the acceleration given. Two of the forces are shown on the free-body diagrams below, but the third is missing. For each, draw and label *on the grid* the missing third force vector.

13. $a_x = 2 \text{ m/s}^2$
$a_y = 0 \text{ m/s}^2$

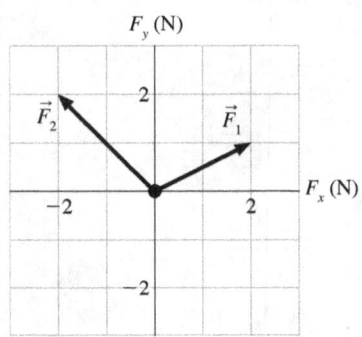

14. $a_x = 0 \text{ m/s}^2$
$a_y = -3 \text{ m/s}^2$

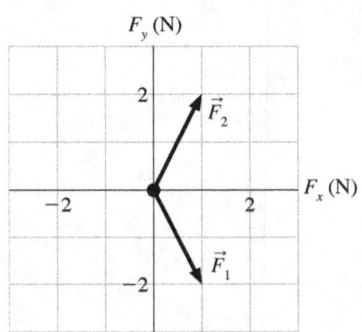

15. The object moves with constant velocity.

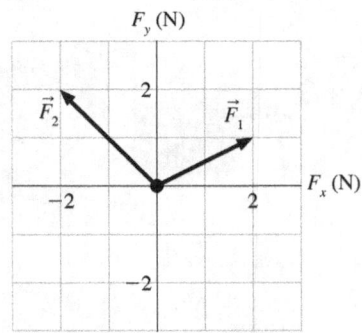

5.3 Mass and Weight

16. Suppose you have a jet-powered flying platform that can move straight up and down. For each of the following cases, is your apparent weight equal to, greater than, or less than your true weight? Explain.

    a. You are ascending and speeding up.

    b. You are descending and speeding up.

    c. You are ascending at a constant speed.

    d. You are ascending and slowing down.

    e. You are descending and slowing down.

17. The terms "vertical" and "horizontal" are frequently used in physics. Give *operational definitions* for these two terms. An operational definition defines a term by how it is measured or determined. Your definition should apply equally well in a laboratory or on a steep mountainside.

18. An astronaut orbiting the earth is handed two balls that are identical in outward appearance. However, one is hollow while the other is filled with lead. How might the astronaut determine which is which? Cutting them open is not allowed.

5.4 Normal Forces

19. Suppose you stand on a spring scale in six identical elevators. Each elevator moves as shown below. Let the reading of the scale in elevator n be S_n. Rank in order, from largest to smallest, the six scale readings S_1 to S_6. Some may be equal. Give your answer in the form $A > B = C > D$.

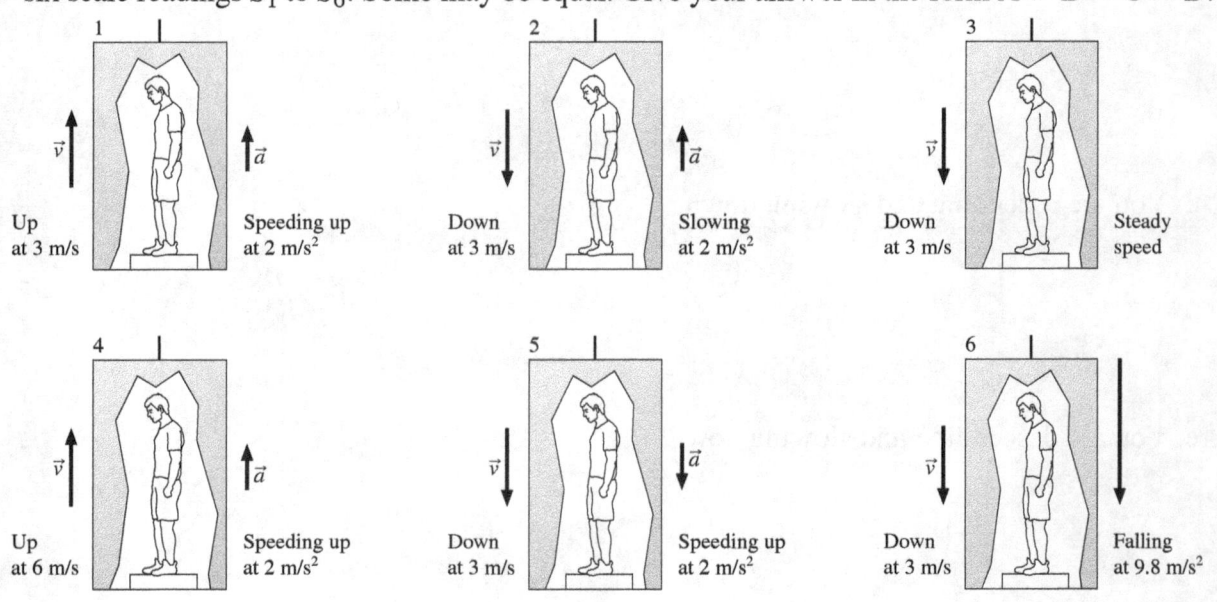

Order:

Explanation:

5.5 Friction

20. A block pushed along the floor with velocity $\vec{v}_0$ slides a distance d after the pushing force is removed.

    a. If the mass of the block is doubled but the initial velocity is not changed, what is the distance the block slides before stopping? Explain.

    b. If the initial velocity of the block is doubled to $2\vec{v}_0$ but the mass is not changed, what is the distance the block slides before stopping? Explain.

21. Consider a box in the back of a pickup truck.

    a. If the truck accelerates slowly, the box moves with the truck without slipping. What force acts on the box to accelerate it? In what direction does this force point?

    b. Draw a free-body diagram of the box.

    c. What happens to the box if the truck accelerates too rapidly? Explain why this happens, basing your explanation on the physical laws and models described in this chapter.

22. A small airplane of mass m must take off from a primitive jungle airstrip that slopes upward at
PSA a slight angle θ. When the pilot pulls back on the throttle, the plane's engines exert a constant
5.2 forward force $\vec{F}_{thrust}$. Rolling friction is not negligible on the dirt airstrip, and the coefficient
of rolling resistance is μ_r. If the plane's take-off speed is v_{off}, what minimum length must the
airstrip have for the plane to get airborne?

a. Assume the plane takes off uphill to the right. Begin with a pictorial representation, as
was described in Tactics Box 2.2. Establish a coordinate system with a tilted x-axis; show
the plane at the beginning and end of the motion; define symbols for position, velocity,
and time at these two points (six symbols all together); list known information; and state
what you wish to find. $\vec{F}_{thrust}$, m, θ, μ_r, and v_{off} are presumed known, although we have only
symbols for them rather than numerical values, and three other quantities are zero.

b. Next, draw a force-identification diagram. Beside it, draw a free-body diagram. Your
free-body diagram should use the same coordinate system you established in part a, and it
should have four forces shown on it.

c. Write Newton's second law as two equations, one for the net force in the x-direction and
one for the net force in the y-direction. Be careful finding the components of $\vec{w}$ (see Figure
5.11), and pay close attention to signs. Remember that symbols such as w or f_k represent
the *magnitudes* of vectors; you have to supply appropriate signs to indicate which way the
vectors point. The right side of these equations have a_x and a_y. The motion is entirely along
the x-axis, so what do you know about a_y? Use this information as you write the y-equation.

d. Now write the equation that characterizes the friction force on a rolling tire.

e. Combine your friction equation with the y-equation of Newton's second law to find an expression for the magnitude of the friction force.

f. Finally, substitute your answer to part e into the x-equation of Newton's second law, and then solve for a_x, the x-component of acceleration. Use $w = mg$ if you've not already done so.

g. With friction present, should the *magnitude* of the acceleration be larger or smaller than the acceleration of taking off on a frictionless runway? _____

h. Does your expression for acceleration agree with your answer to part g? _____
   Explain how you can tell. If it doesn't, recheck your work.

i. The force analysis is done, but you still have to do the kinematics. This is a situation where we know about velocities, distance, and acceleration but nothing about the time involved. That should suggest the appropriate kinematics equation. Use your acceleration from part f in that kinematics equation, and solve for the unknown quantity you're seeking.

You've found a symbolic answer to the problem, one that you could now evaluate for a range of values of $\vec{F}_{\text{thrust}}$ or θ without having to go through the entire solution each time.

You Write the Problem!

Exercises 23–25: You are given the dynamics equations that are used to solve a problem. For each of these:

a. Write a *realistic* physics problem for which these are the correct equations. Look at worked examples and end-of-chapter problems in the textbook to see what realistic physics problems are like. Be sure that the problem you write, and the answer you ask for, is consistent with the information given in the equation.

b. Draw the free-body diagram and pictorial representation for your problem.

c. Finish the solution of the problem.

23. $-0.80n = (1500 \text{ kg})a_x$

$n - (1500 \text{ kg})(9.8 \text{ m/s}^2) = 0$

24. $T - 0.20n - (20 \text{ kg})(9.8 \text{ m/s}^2)\sin 20° = (20 \text{ kg})(2.0 \text{ m/s}^2)$

$n - (20 \text{ kg})(9.8 \text{ m/s}^2)\cos 20° = 0$

25. $(100 \text{ N})\cos 30° - f_k = (20 \text{ kg})a_x$

$n + (100 \text{ N})\sin 30° - (20 \text{ kg})(9.8 \text{ m/s}^2) = 0$

$f_k = 0.20n$

5.6 Drag

26. Five balls move through the air as shown. All five have the same size and shape. Rank in order, from largest to smallest, the magnitude of their accelerations a_1 to a_4. Some may be equal. Give your answer in the form $A > B = C > D$.

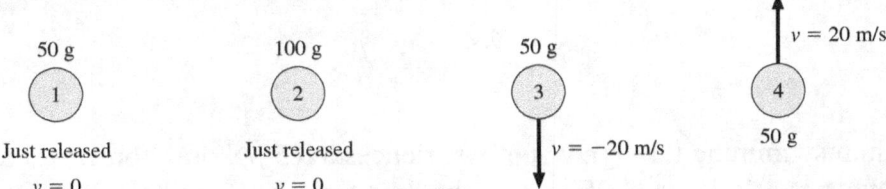

Order:

Explanation:

27. A 1 kg wood ball and a 5 kg metal ball have identical shapes and sizes. They are dropped simultaneously from a tall tower. Each ball experiences the *same* drag force because both have the same shape.

a. Draw free-body diagrams for the two balls as they fall in the presence of air resistance. Make sure your vectors all have the correct *relative* lengths.

b. Both balls have the same acceleration if they fall in a vacuum. The metal ball has a larger weight, but it also has a larger mass and so $a = F/m$ is the same as for the wood ball. But the accelerations are not the same if the balls fall in air. Which now has the larger acceleration? Your explanation should refer explicitly to your free-body diagrams of part a and to Newton's second law.

c. Do the balls hit the ground simultaneously? If not, which hits first? Explain.

28. a. A ball sailing through the air experiences a 0.5 N drag force. What will the drag force be if the ball is thrown twice as fast? Explain.

   b. A bacterium swimming through water experiences a 0.5 pN drag force, where 1 pN = 1 piconewton = 1×10^{-12} N. What will the drag force be if the bacterium swims twice as fast? Explain.

5.7 Interacting Objects

29. Block A is pushed across a horizontal surface at a *constant* speed by a hand that exerts force $\vec{F}_{\text{H on A}}$. The surface has friction.

    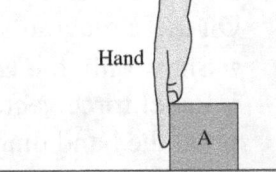

    a. Draw two free-body diagrams, side by side, one for the hand and the other for the block. On these diagrams:
       - Show only the *horizontal* forces on each object.
       - Label force vectors using the form $\vec{F}_{\text{C on D}}$.
       - On the hand diagram show both $\vec{F}_{\text{A on H}}$ and $\vec{F}_{\text{arm on H}}$.
       - Make sure vector lengths correctly indicate the relative magnitudes of the forces.

    b. Rank in order, from largest to smallest, the magnitudes of *all* of the horizontal forces you showed in part a. For example, if $\vec{F}_{\text{C on D}}$ is the largest of three forces while $\vec{F}_{\text{D on C}}$ and $\vec{F}_{\text{D on E}}$ are smaller but equal, you can record this as $\vec{F}_{\text{C on D}} > \vec{F}_{\text{D on C}} = \vec{F}_{\text{D on E}}$.

    Order:

    Explanation:

30. A second block B is placed in front of Block A of question 28. B is more massive than A: $m_{\text{B}} > m_{\text{A}}$. The blocks are speeding up.

    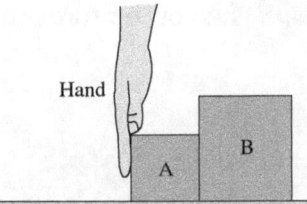

    a. Consider a *frictionless* surface. Draw separate free-body diagrams for A, B, and the hand. Show only the horizontal forces. Label forces in the form $\vec{F}_{\text{C on D}}$.

    b. By applying Newton's second law to each block and the third law to each action/reaction pair, rank in order *all* of the horizontal forces, from largest to smallest.

    Order:

    Explanation:

31. Blocks A and B are held on the palm of your outstretched hand as you lift them straight up at *constant speed*. Assume $m_B > m_A$ and that $m_{hand} = 0$.

    a. Draw three free-body diagrams, stacked vertically, for A, B, and your hand. On these diagrams:
       - Show only the *vertical* forces on each object, including the weights of the blocks.
       - Label force vectors using the form $\vec{F}_{C\ on\ D}$.
       - On the hand diagram show both $\vec{F}_{A\ on\ H}$ and $\vec{F}_{arm\ on\ H}$.
       - Make sure vector lengths correctly indicate the relative magnitudes of the forces.

    b. Rank in order, from largest to smallest, all of the vertical forces. Explain your reasoning.

32. A mosquito collides head-on with a car traveling 60 mph.

    a. How do you think the size of the force that the car exerts on the mosquito compares to the size of the force that the mosquito exerts on the car?

    b. Draw *separate* free-body diagrams of the car and the mosquito at the moment of collision, showing only the horizontal forces. Label forces in the form $\vec{F}_{C\ on\ D}$.

    c. Does your answer to part b confirm your answer to part a? Explain why or why not.

5.8 Ropes and Pulleys

33. Blocks A and B are connected by a massless string over a massless, frictionless pulley. The blocks have just this instant been released from rest.

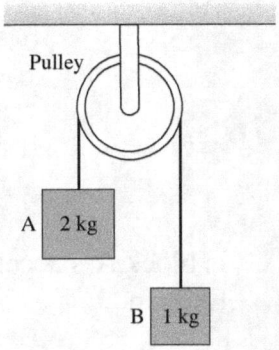

a. Will the blocks accelerate? If so, in which directions?

b. Draw a separate free-body diagram for each block. Be sure vector lengths indicate the relative sizes of the forces.

c. Rank in order, from largest to smallest, all of the vertical forces. Explain.

d. Consider the block that falls. Is the magnitude of its acceleration less than, greater than, or equal to g? Explain.

34. In case a, block A is accelerated across a frictionless table by a hanging 10 N weight (1.02 kg). In case b, the same block is accelerated by a steady 10 N tension in the string.

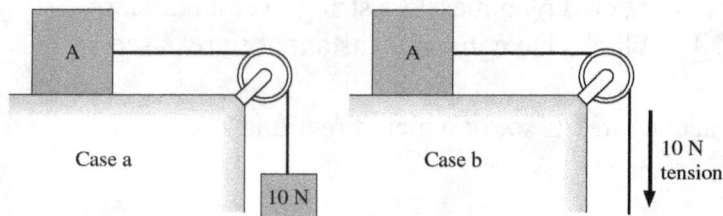

Is block A's acceleration in case b greater than, less than, or equal to its acceleration in case a? Explain.

Exercises 34–35: Draw separate free-body diagrams for blocks A and B. Indicate any pairs of forces having the same magnitude.

35.

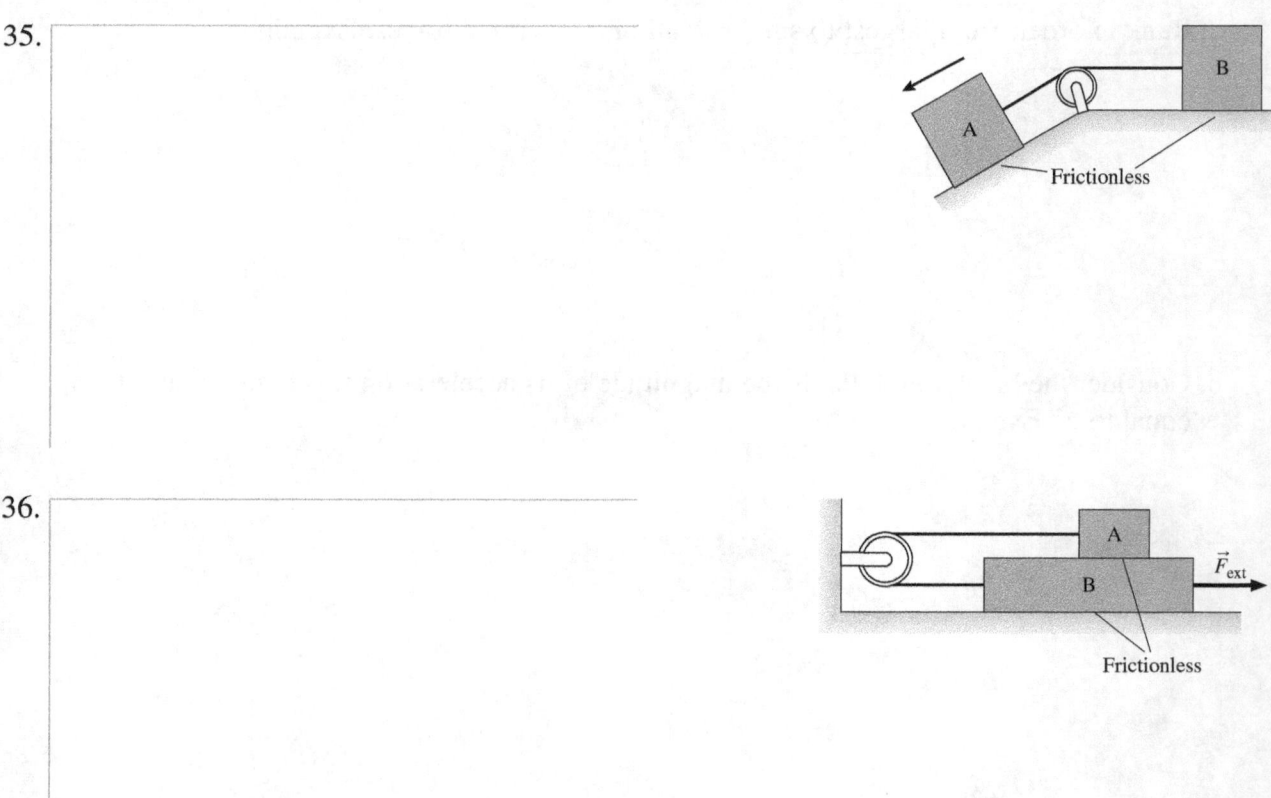

36.

6 Circular Motion, Orbits, and Gravity

6.1 Uniform Circular Motion

1. a. The crankshaft in your car rotates at 3000 rpm. What is the frequency in revolutions per second?

   b. A record turntable rotates at 33.3 rpm. What is the rotation period in seconds?

2. The figure shows three points on a steadily rotating wheel.

   a. Draw the velocity vectors at each of the three points.
   b. Rank in order, from largest to smallest, the speeds v_1, v_2, and v_3 of these points.

   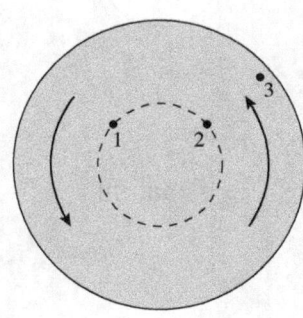

   Order:

   Explanation:

3. An object in uniform circular motion has acceleration $a = 8$ m/s^2. What is a if

   a. The circle's radius is doubled without changing the frequency? _____

   b. The radius is doubled without changing the object's speed? _____

   c. The frequency is doubled without changing the circle's radius? _____

6.2 Dynamics of Uniform Circular Motion

4. The figure shows a *top view* of a plastic tube that is fixed on a horizontal table top. A marble is shot into the tube at A. Sketch the marble's trajectory after it leaves the tube at B. Explain.

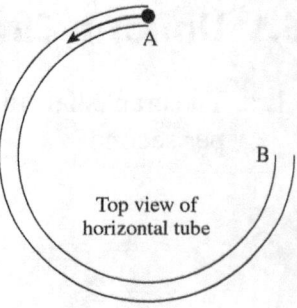

Top view of horizontal tube

5. The figures are a bird's-eye view of particles moving in horizontal circles on a table top. All are moving at the same speed. Rank in order, from largest to smallest, the tensions T_1 to T_4.

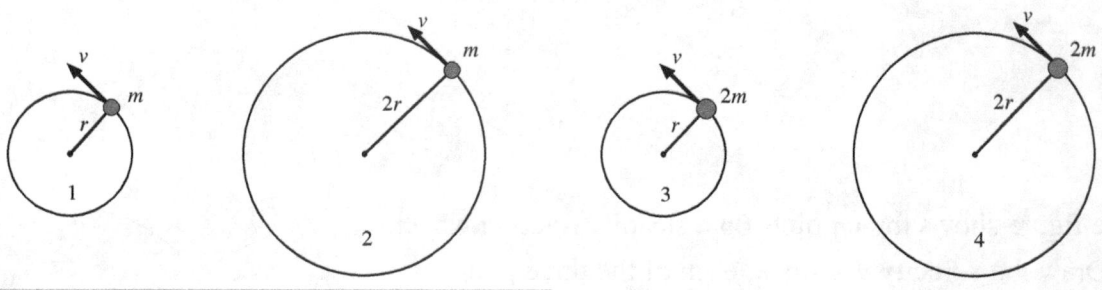

Order:

Explanation:

6. A ball rolls over the top of a circular hill. Rolling friction is negligible. Circle the letter of the ball's free-body diagram at the very top of the hill.

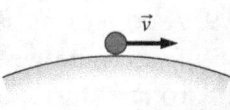

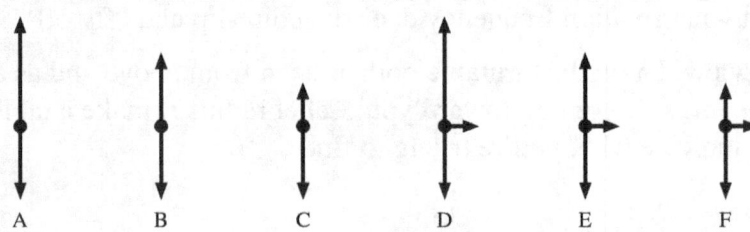

A   B   C   D   E   F

Explanation:

7. A ball on a string moves in a vertical circle. When the ball is at its lowest point, is the tension in the string greater than, less than, or equal to the ball's weight? Explain. (You should include a free-body diagram as part of your explanation.)

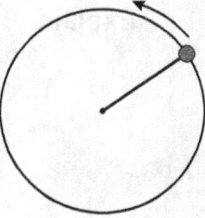

8. A marble rolls around the inside of a cone. Draw a free-body diagram of the marble when it is on the left side of the cone, coming toward you.

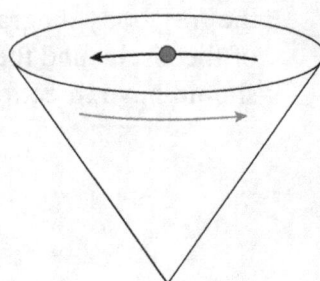

9. A coin of mass m is placed at distance r from the center of a turntable. The coefficient of static
PSA friction between the coin and the turntable is μ_s. Starting from rest, the turntable is gradually
6.1 rotated faster and faster. At what rotation frequency does the coin slip and "fly off"?

a. Begin with a visual overview. Draw the turntable both as seen from above and as an edge view with the coin on the left side coming toward you. Label radius r, make a table of known information, and indicate what you're trying to find.

b. What direction does $\vec{f}_s$ point? Explain.

c. What condition describes the situation just as the coin starts to slip? Write this condition as a mathematical statement.

d. Now draw a free-body diagram of the coin. Following Problem Solving Strategy 6.1, draw the free-body diagram with the circle viewed edge on, the x-axis pointing toward the center of the circle, and the y-axis perpendicular to the plane of the circle. Your free-body diagram should have three forces on it.

e. Referring to Problem Solving Strategy 6.1, write Newton's second law for the x- and y-components of the forces. One sum should equal 0, the other mv^2/r.

f. The two equations of part e are valid for any speed up to the point of slipping. If you combine these with your statement of part c, you can solve for the speed v_{max} at which the coin slips. Do so.

g. Finally, use the relationship between v and f to find the frequency at which the coin slips.

6.3 Apparent Forces in Circular Motion

10. The drawing is a partial motion diagram for a car rolling at constant speed over the top of a circular hill.

    a. Complete the motion diagram by adding the car's velocity vectors. Then use the velocity vectors to determine *and show* the car's acceleration $\vec{a}$ at the top of the hill.
    b. To the right, draw a free-body diagram of the car at the top of the hill. Next to the free-body diagram, indicate the direction of the net force on the car or, if appropriate, write $\vec{F}_{net} = \vec{0}$.

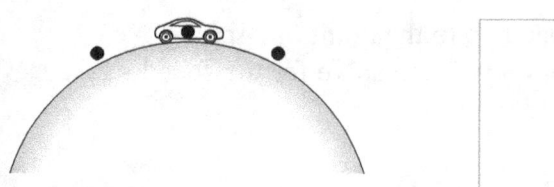

    c. Does the net force point of your free-body diagram point in the same direction you showed for the car's acceleration? _____
    If not, you may want to reconsider your work this far because Newton's second law requires $\vec{F}_{net}$ and $\vec{a}$ to point the same way.
    d. Is there a maximum speed at which the car can travel over the top of the hill and not lose contact with the ground? If so, show how the free-body diagram would look at that speed. If not, why not?

11. A stunt plane does a series of vertical loop-the-loops. At what point in the circle does the pilot feel the heaviest? Explain. Include a free-body diagram with your explanation.

12. A roller-coaster car goes around the inside of a loop-the-loop. One of the following statements is true at the highest point in the loop, and one is true at the lowest point. Check the true statements.

	Highest	Lowest
The car's apparent weight w_{app} is always less than w		
The car's apparent weight w_{app} is always equal to w		
The car's apparent weight w_{app} is always greater than w		
w_{app} could be less than, equal to, or greater than w		

13. You can swing a ball on a string in a *vertical* circle without the string going slack if you swing it fast enough.

   a. Draw two free-body diagrams of the ball at the top of the circle. On the left, show the ball when it is going around the circle very fast. On the right, show the ball as it goes around the circle more slowly.

   Very fast

   Slower

   b. Suppose the ball has the smallest possible frequency that allows it to go all the way around the circle. What is the tension in the string when the ball is at the highest point? Explain.

6.4 Circular Orbits and Weightlessness

14. The Earth has seasons because the axis of the Earth's rotation is tilted 23° away from a line perpendicular to the plane of the Earth's orbit. You can see this in the figure, which shows the edge of the Earth's orbit around the sun. For both positions of the Earth, draw a force vector to show the net force acting on the Earth or, if appropriate, write $\vec{F} = \vec{0}$.

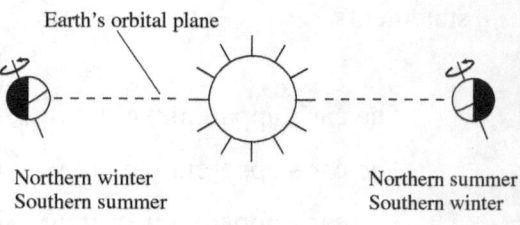

15. A small projectile is launched parallel to the ground at height $h = 1$ m with sufficient speed to orbit a completely smooth, airless planet. A bug rides in a small hole inside the projectile. Is the bug weightless? Explain.

16. It's been proposed that future space stations create "artificial gravity" by rotating around an axis. (The space station would have to be much larger than the present space station for this to be feasible.)

    a. How would this work? Explain.

    b. Would the artificial gravity be equally effective throughout the space station? If not, where in the space station would the residents want to live and work?

6.5 Newton's Law of Gravity

17. Is the Earth's gravitational force on the sun larger than, smaller than, or equal to the sun's gravitational force on the Earth? Explain.

18. Star A is twice as massive as star B.

    a. Draw gravitational force vectors on both stars. The length of each vector should be proportional to the size of the force.

    A $m_A = 2m_B$ B m_B

    b. Is the acceleration of star A toward B larger than, smaller than, or equal to the acceleration of star B toward A? Explain.

19. The quantity y is inversely proportional to the square of x, and $y = 4$ when $x = 5$.

    a. Write an equation to represent this inverse-square relationship for all y and x.

    b. Find y if $x = 2$. _____   c. Find x if $y = 100$. _____

20. The quantity y is inversely proportional to the square of x. For one value of x, $y = 12$.

    a. What is the value of y if x is doubled? _____

    b. What is the value of y if the original value of x is halved? _____

21. How far away from the Earth does an orbiting spacecraft have to be in order for the astronauts inside to be "weightless"?

22. The acceleration due to gravity at the surface of Planet X is 20 m/s^2. The radius and the mass of Planet Z are twice those of Planet X. What is g on Planet Z?

6.6 Gravity and Orbits

23. Planet X orbits the star Omega with a "year" that is 200 Earth days long. Planet Y circles Omega at four times the distance of Planet X. How long is a year on Planet Y?

24. The mass of Jupiter is $M_{\text{Jupiter}} = 300 M_{\text{Earth}}$. Jupiter orbits around the Sun with $T_{\text{Jupiter}} = 11.9$ years in an orbit with $r_{\text{Jupiter}} = 5.2 r_{\text{Earth}}$. Suppose the Earth could be moved to the distance of Jupiter and placed in a circular orbit around the Sun. The new period of the Earth's orbit would be

a. 1 year.

b. 11.9 years.

c. Between 1 year and 11.9 years.

d. More than 11.9 years.

e. It could be anything, depending on the speed the Earth is given.

f. It is impossible for a planet of Earth's mass to orbit at the distance of Jupiter.

Circle the letter of the true statement. Then explain your choice.

25. The gravitational force of a star on orbiting planet 1 is F_1. Planet 2, which is twice as massive as Planet 1 and orbits at twice the distance from the star, experiences gravitational force F_2.

    a. What is the ratio F_2/F_1?

    b. Planet 1 orbits the star with period T_1 and Planet 2 with period T_2. What is the ratio T_2/T_1?

26. Satellite A orbits a planet with a speed of 10,000 m/s. Satellite B, orbiting at the same distance from the center of the planet, is twice as massive as Satellite A. What is the speed of Satellite B?

You Write the Problem!

Exercises 27–30: You are given the equation that is used to solve a problem. For each of these:

a. Write a realistic physics problem for which this is the correct equation. Look at worked examples and end-of-chapter problems in the textbook to see what realistic physics problems are like. Be sure that the problem you write, and the answer you ask for, is consistent with the information given in the equation.

b. If appropriate, draw the free-body diagram and pictorial representation for your problem.

c. Finish the solution of the problem.

27. $60 \text{ N} = \dfrac{(0.30 \text{ kg}) v^2}{0.50 \text{ m}}$

28. $(1500 \text{ kg})(9.8 \text{ m/s}^2) - 11{,}760 \text{ N} = \dfrac{(1500 \text{ kg}) v^2}{200 \text{ m}}$

Note: See the Astronomical Data table inside the back cover of your textbook to help interpret the equations in Exercises 29 and 30.

29. $\dfrac{(6.67 \times 10^{-11} \text{ N} \cdot \text{m}^2/\text{kg}^2)(1.90 \times 10^{27} \text{ kg})}{R^2} = \dfrac{(6.67 \times 10^{-11} \text{ N} \cdot \text{m}^2/\text{kg}^2)(5.98 \times 10^{24} \text{ kg})}{6.37 \times 10^6 \text{ m}}$

30. $\dfrac{(6.67 \times 10^{-11} \text{ N} \cdot \text{m}^2/\text{kg}^2)(5.98 \times 10^{24} \text{ kg})(1000 \text{ kg})}{r^2} = \dfrac{(1000 \text{ kg})(1997 \text{ m/s})^2}{r}$

7 Rotational Motion

7.1 Describing Circular and Rotational Motion

1. A particle undergoes uniform circular motion with constant angular velocity $\omega = +1.0$ rad/s, starting from point P.

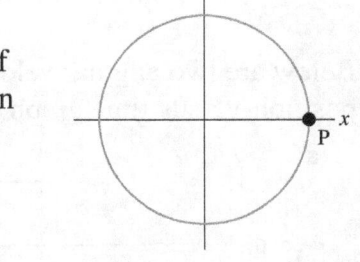

   a. On the figure, draw a motion diagram showing the location of the particle every 1.0 s until the particle has moved through an angle of 5 rad. Draw velocity vectors **black** and acceleration vectors **red**. For this question, you can use 1 rad $\approx 60°$.

   b. Below, graph the particle's angular position θ and angular velocity ω for the first 5 s of motion. Include an appropriate vertical scale on both graphs.

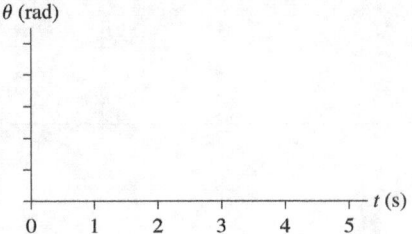

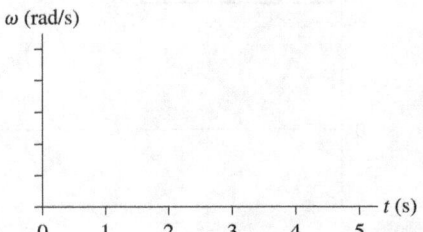

2. Point B on a rotating wheel is twice as far from the axle as point A.

   a. Is ω_B equal to $\frac{1}{2}\omega_A$, ω_A, or $2\omega_A$? Explain.

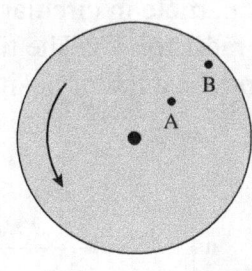

   b. Is v_B equal to $\frac{1}{2}v_A$, v_A, or $2v_A$? Explain.

3. Below are two angular position-versus-time graphs. For each, draw the corresponding angular velocity-versus-time graph directly below it.

a.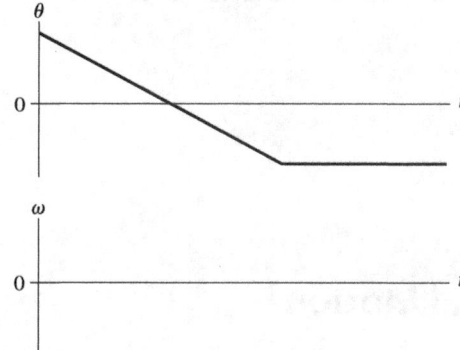

b.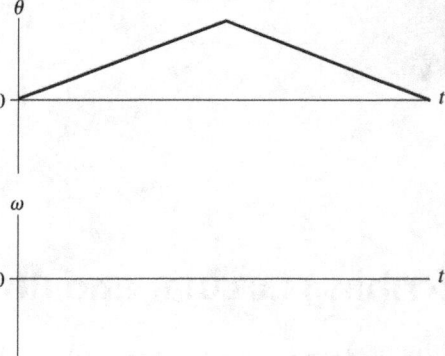

4. Below are two angular velocity-versus-time graphs. For each, draw the corresponding angular position-versus-time graph directly below it. Assume $\theta_0 = 0$ rad.

a.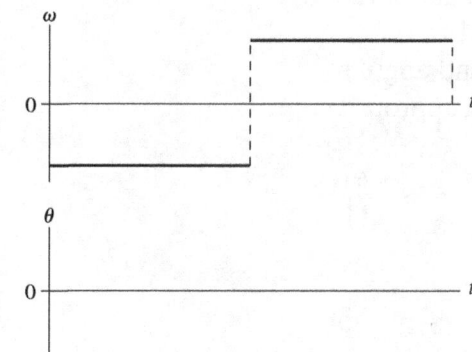

b.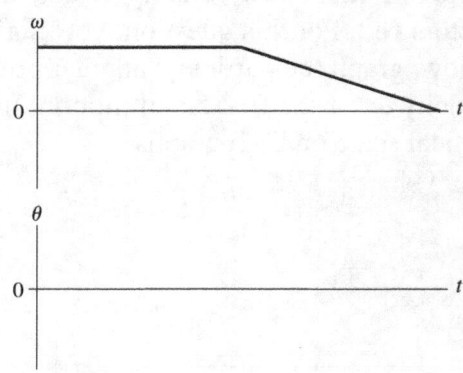

5. A particle in circular motion rotates clockwise at 4 rad/s for 2 s, and then counterclockwise at 2 rad/s for 4 s. The time required to change direction is negligible. Graph the angular velocity and the angular position, assuming $\theta_0 = 0$ rad.

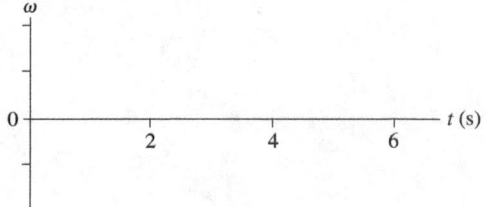

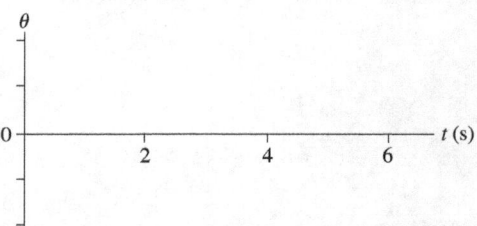

7.2 The Rotation of a Rigid Body

6. The following figures show a rotating wheel. If we consider a counterclockwise rotation as positive (+) and a clockwise rotation as negative (−), determine the signs (+ or −) of ω and α.

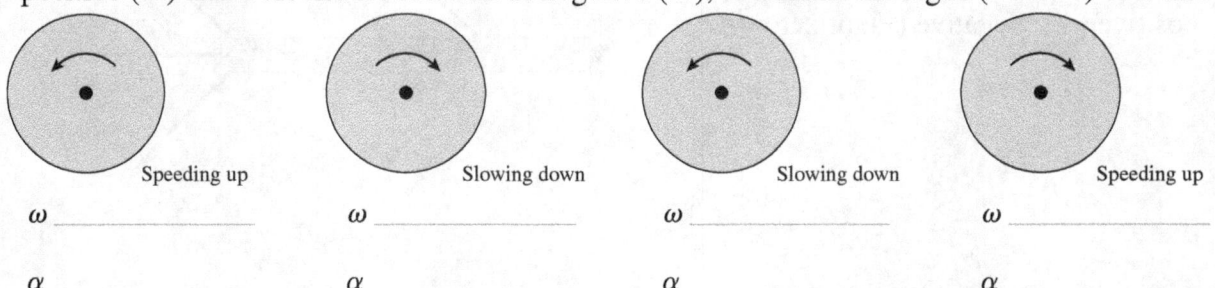

Speeding up   Slowing down   Slowing down   Speeding up

ω _____ ω _____ ω _____ ω _____

α _____ α _____ α _____ α _____

7. A ball is rolling back and forth inside a bowl. The figure shows the ball at extreme left edge of the ball's motion as it changes direction.

   a. At this point, is ω positive, negative, or zero? _____

   b. At this point, is α positive, negative, or zero? _____

8. The figures below show the centripetal acceleration vector $\vec{a}_c$ at four successive points on the trajectory of a particle moving in a counterclockwise circle.

   a. For each, draw the tangential acceleration vector $\vec{a}_t$ at points 2 and 3 or, if appropriate, write $\vec{a}_t = \vec{0}$.

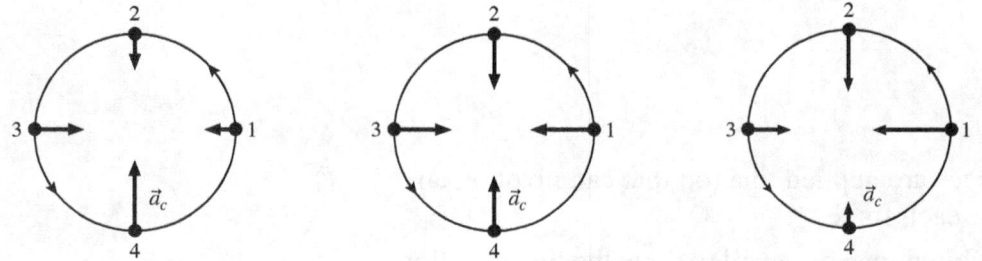

   b. If we consider a counterclockwise rotation as positive and a clockwise rotation as negative, determine if the particle's angular acceleration α is positive (+), negative (−), or zero (0).

   $\alpha =$ _____ $\alpha =$ _____ $\alpha =$ _____

9. Below are three angular velocity-versus-time graphs. For each, draw the corresponding angular acceleration-versus-time graph.

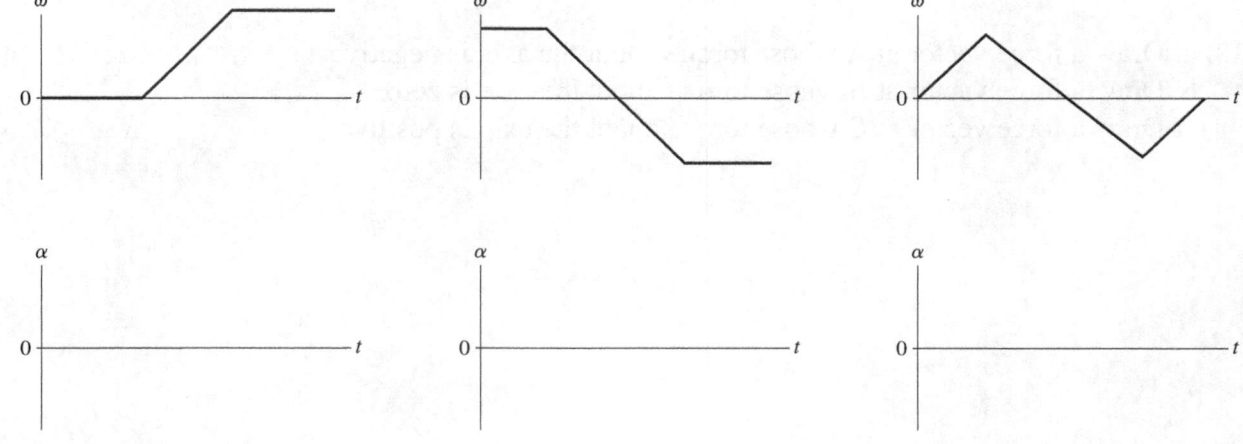

7.3 Torque

10. Five forces are applied to a door. For each, determine if the torque about the hinge is positive (+), negative (−), or zero (0).

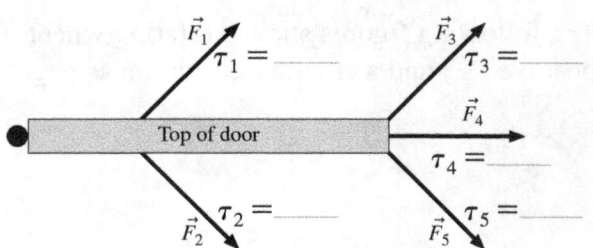

11. Four forces, each of magnitude either F or $2F$, are applied to a door. Rank in order, from largest to smallest, the four torques τ_1 to τ_4 about the hinge.

Order:

Explanation:

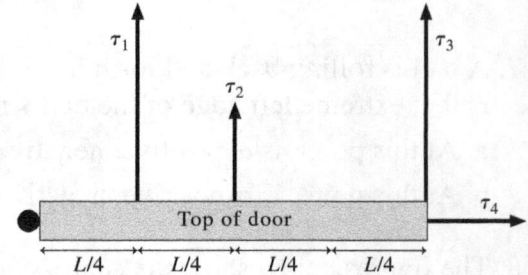

12. Four forces are applied to a rod that can pivot on an axle. For each force,

   a. Use a **black** pen or pencil to draw the line of action.

   b. Use a **red** pen or pencil to draw and label the moment arm, or state that $r_\perp = 0$.

   c. Determine if the torque about the axle is positive (+), negative (−), or zero (0).

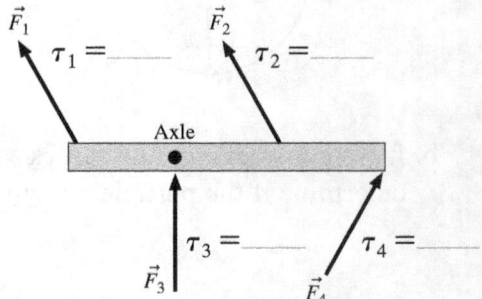

13. a. Draw a force vector at A whose torque about the axle is negative.
   b. Draw a force vector at B whose torque about the axle is zero.
   c. Draw a force vector at C whose torque about the axle is positive.

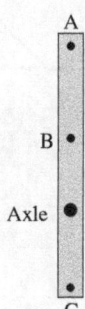

7.4 Gravitational Torque and the Center of Gravity

14. Two spheres are connected by a rigid but massless rod to form a dumbbell. The two spheres have equal mass. Mark the approximate location of the center of gravity with an ×. Explain your reasoning.

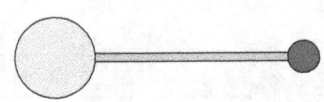

15. Two spheres are connected by a rigid but massless rod to form a dumbbell. The two spheres are made of the same material. Mark the approximate location of the center of gravity with an ×. Explain your reasoning.

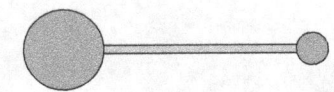

16. Mark the center of gravity of this object with an ×. Explain.

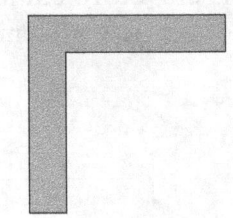

17. a. Find the coordinate for and *mark* the center of gravity for the pair of masses shown, using the center of the 3.0 kg mass as the origin.

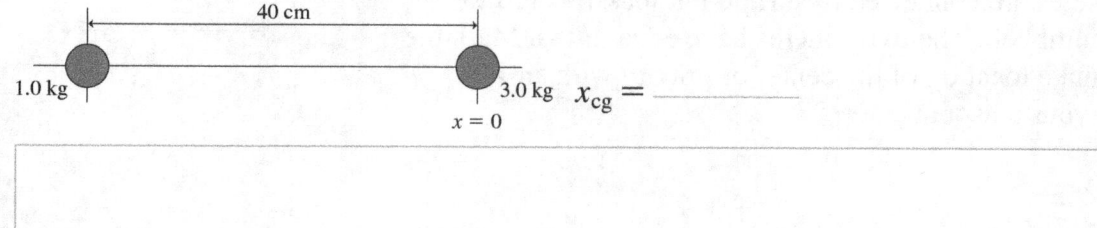

b. Find the coordinate for and *mark* the center of gravity for the pair of masses shown, using the center of the 1.0 kg mass as the origin.

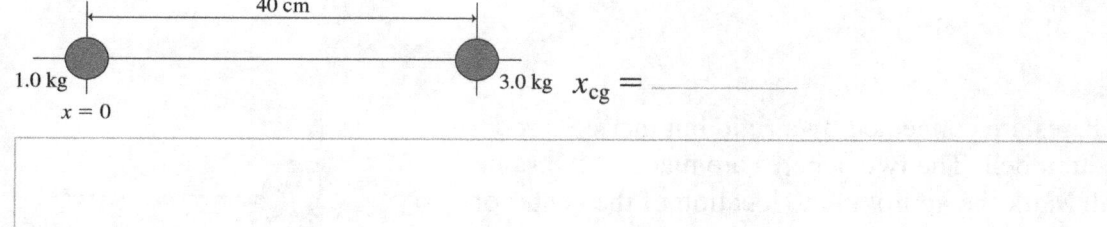

c. How do the locations for the marks in parts a and b compare? How do the coordinates compare?

d. The 3.0 kg mass from parts a and b above is separated into two 1.5 kg pieces. One of these is moved 10 cm in the $+y$-direction. Find the coordinates for and *mark* the center of gravity of the new system using the center of the 1.0 kg mass as the origin.

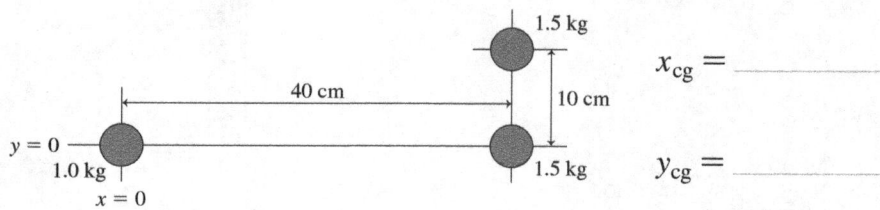

e. What effect did separating the 3.0 kg mass along the y-direction have on the x-component of the center of gravity of the system?

7.5 Rotational Dynamics and Moment of Inertia

18. A student gives a quick push to a ball at the end of a massless, rigid rod, causing the ball to rotate clockwise in a *horizontal* circle. The rod's pivot is frictionless.

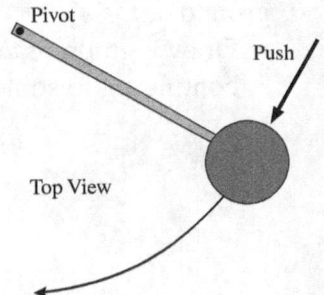

    a. As the student is pushing, is the torque about the pivot positive, negative, or zero?_____

    b. After the push has ended, does the ball's angular velocity

       i.   Steadily increase?

       ii. Increase for awhile, then hold steady?

       iii. Hold steady?

       iv. Decrease for awhile, then hold steady?

       v.   Steadily decrease?

       Explain the reason for your choice.

    c. Right after the push has ended, is the torque positive, negative, or zero?_____

19. The top graph shows the torque on a rotating wheel as a function of time. The wheel's moment of inertia is $10 \text{ kg} \cdot \text{m}^2$. Draw graphs of α-versus-t and ω-versus-t, assuming $\omega_0 = 0$. Provide units and an appropriate scale on the vertical axis.

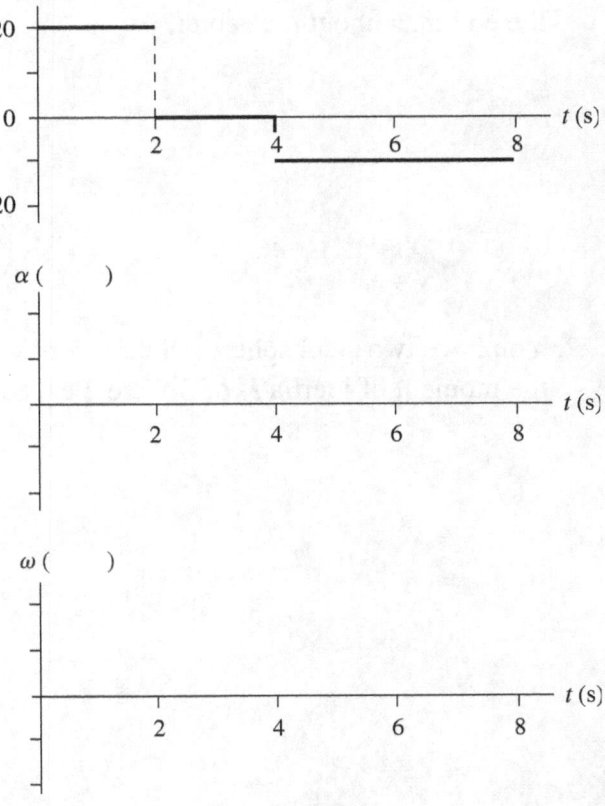

20. The wheel turns on a frictionless axle. A string wrapped around the smaller diameter shaft is tied to a block. The block is released at $t = 0$ s and hits the ground at $t = t_1$.

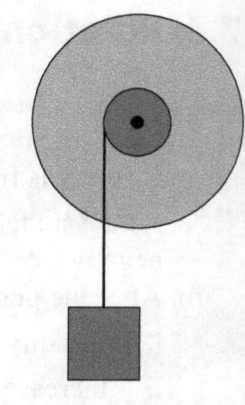

   a. Draw a graph of ω-versus-t for the wheel, starting at $t = 0$ s and continuing to some time $t > t_1$.

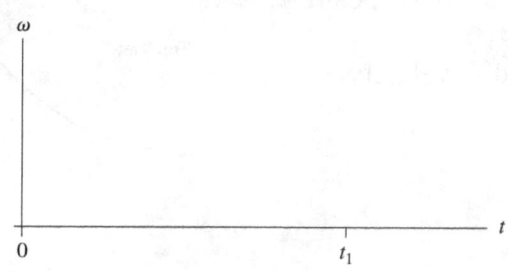

   b. Is the magnitude of the block's downward acceleration greater than g, less than g, or equal to g? Explain.

21. The moment of inertia of a uniform rod about an axis through its center is $\frac{1}{12}ML^2$. The moment of inertia about an axis at one end is $\frac{1}{3}ML^2$. Explain *why* the moment of inertia is larger about the end than about the center.

22. You have two steel spheres. Sphere 2 has twice the radius of Sphere 1. By what *factor* does the moment of inertia I_2 of Sphere 2 exceed the moment of inertia I_1 of Sphere 1?

23. The professor hands you two spheres. They have the same mass, the same radius, and the same exterior surface. The professor claims that one is a solid sphere and that the other is hollow. Can you determine which is which without cutting them open? If so, how? If not, why not?

24. Rank in order, from largest to smallest, the moments of inertia I_1, I_2, and I_3 about the midpoint of each connecting rod.

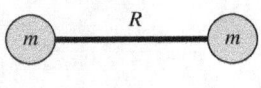

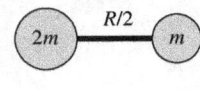

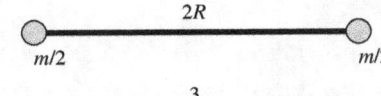

1   2   3

Order:

Explanation:

7.6 Using Newton's Second Law for Rotation

25. A square plate can rotate about an axle through its center. Four forces of equal magnitude are applied to different points on the plate. The forces turn as the plate rotates, maintaining the same orientation with respect to the plate. Rank in order, from largest to smallest, the angular accelerations α_1 to α_4.

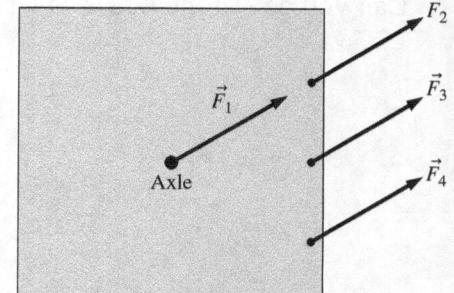

Order:

Explanation:

26. A solid cylinder and a cylindrical shell have the same mass, same radius, and turn on frictionless, horizontal axles. (The cylindrical shell has lightweight spokes connecting the shell to the axle.) A rope is wrapped around each cylinder and tied to a block. The blocks have the same mass and are held at the same height above the ground. Both blocks are released simultaneously. The ropes do not slip.

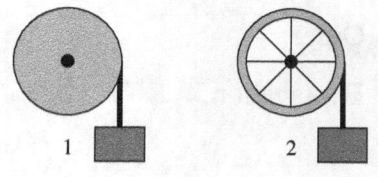

Which block hits the ground first? Or is it a tie? Explain.

27. A metal bar of mass M and length L can rotate in a horizontal
PSA plane about a vertical, frictionless axle through its center. A
7.1 hollow channel down the bar allows compressed air (fed in at
the axle) to spray out of two small holes at the ends of the bar,
as shown. The bar is found to speed up to angular velocity ω
in a time interval Δt, starting from rest. What force does each
escaping jet of air exert on the bar?

Axle
Top view

a. <u>On the figure</u>, draw vectors to show the forces exerted on the
bar. Then label the moment arms of each force.

b. The forces in your drawing exert a torque about the axle.
Write an expression for each torque, and then add them to get the net torque. Your expression should be in terms of the unknown force F and "known" quantities such as M, L, g, etc.

c. What is the moment of inertia of this bar about the axle? _____

d. According to Newton's second law, the torque causes the bar to undergo an angular acceleration. Use your results from parts b and c to write an expression for the angular acceleration. Simplify the expression as much as possible.

e. You can now use rotational kinematics to write an expression for the bar's angular velocity after time Δt has elapsed. Do so.

f. Finally, solve your equation in part e for the unknown force.

This is now a result you could use with experimental measurements to determine the size of the force exerted by the gas.

7.7 Rolling Motion

28. A wheel is rolling along a horizontal surface with the velocity shown. Draw the velocity vectors $\vec{v}_1$ to $\vec{v}_4$ at points 1 to 4 on the rim of the wheel.

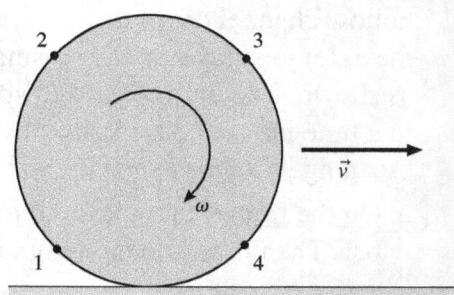

29. A wheel is rolling along a horizontal surface with the velocity shown. Draw the velocity vectors $\vec{v}_1$ to $\vec{v}_3$ at points 1 to 3 on the wheel.

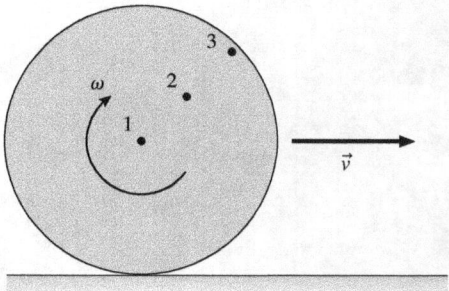

30. If a solid disk and a circular hoop of the same mass and radius are released from rest at the top of a ramp and allowed to roll to the bottom, the disk will get to the bottom first. Explain why this is so.

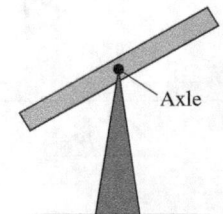
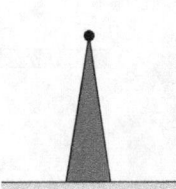

8 Equilibrium and Elasticity

8.1 Torque and Static Equilibrium

1. A uniform rod pivots about a frictionless, horizontal axle through its center. It is placed on a stand, held motionless in the position shown, and then gently released. On the right side of the figure, draw the final, equilibrium position of the rod. Explain your reasoning.

Axle

2. The two objects shown below each have two forces acting on them. The forces have magnitude F or a multiple of F. On each figure, draw a third force that will cause the object to be in static equilibrium. Place the *tail* of the force vector at the point where the force acts. Label your vector with the magnitude of the force as a multiple of F.

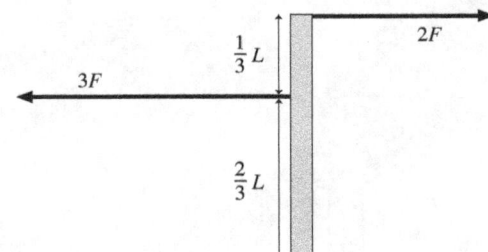
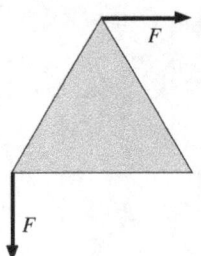

3. A square plate is acted on by the three forces shown. Is the plate in static equilibrium?

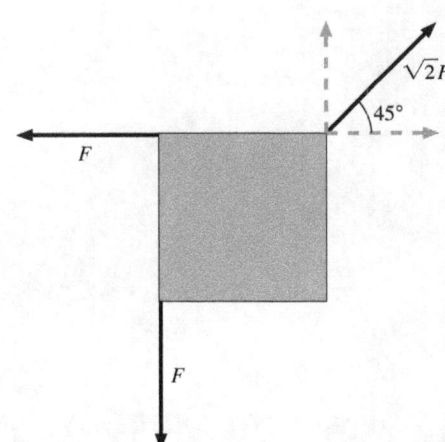

_____ Yes.

_____ No, the net torque is zero but the net force is not zero.

_____ No, the net force is zero but the net torque is not zero.

_____ No, neither the net force nor the net torque is zero.

Exercises 4–7: Each of the following shows a uniform beam of weight w and several blocks, each of which also has weight w. The gravitational force acting downward on the center of gravity of the beam is already shown. For each:

- Draw and label force vectors at each point where a contact force acts on the beam. You should have both upward and downward vectors. The length of each vector should indicate its magnitude relative to the vector showing the weight of the beam.
- Explain why the beam is or is not in equilibrium.

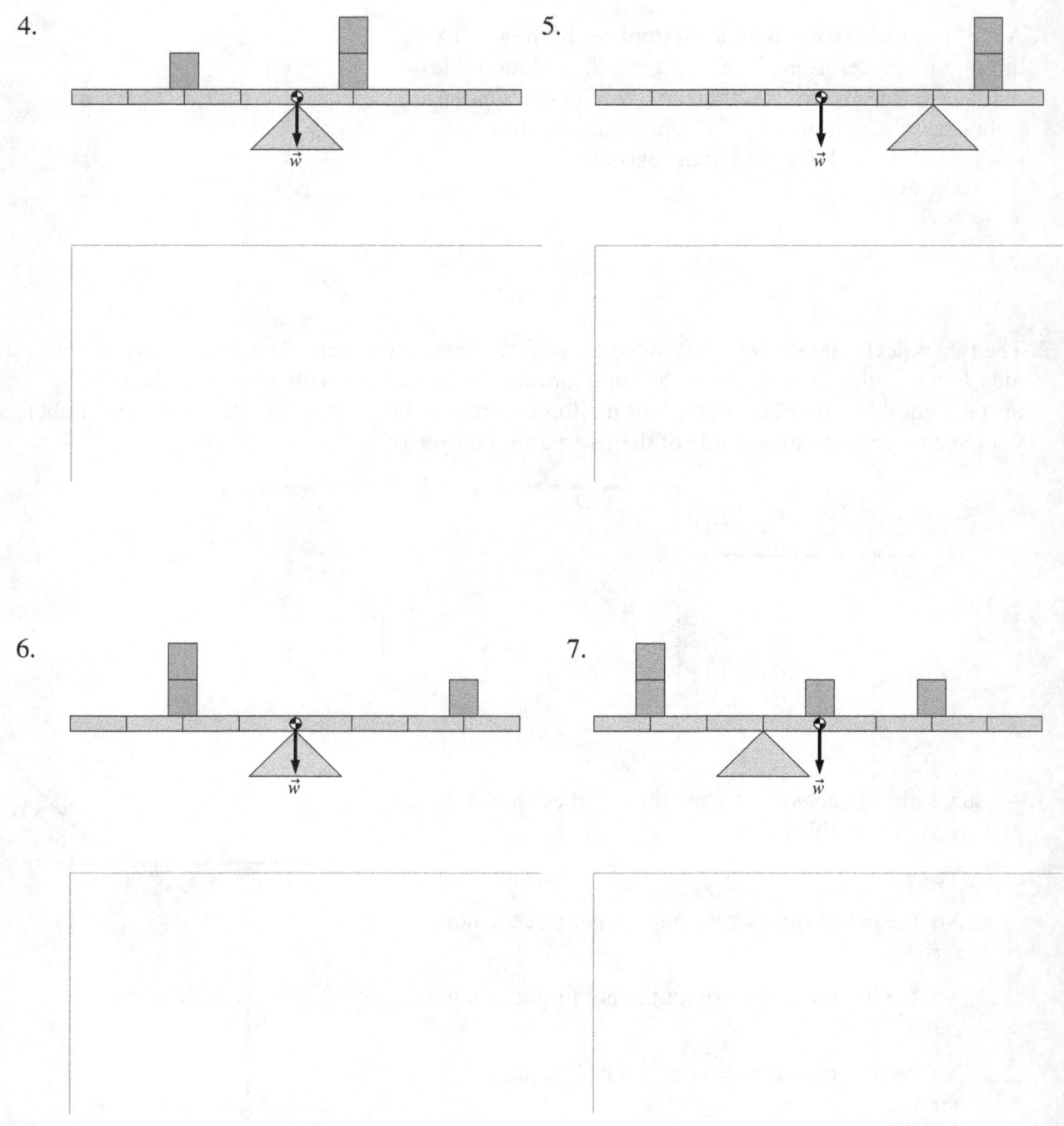

4.

5.

6.

7.

8.2 Stability and Balance

8. The center of gravity is marked on each of the objects below.
   - Show and label with a *t* the object's track width or base of support.
   - Show and label with an *h* the height of the object's center of gravity.
   - Use a ruler to measure the track width and the height of the center of gravity, and then calculate the critical angle.

a.    b.    c.

$t =$ _____ $t =$ _____ $t =$ _____

$h =$ _____ $h =$ _____ $h =$ _____

$\theta_c =$ _____ $\theta_c =$ _____ $\theta_c =$ _____

9. Tightrope walkers often carry a long pole that is weighted at the end. The pole serves two purposes: to change the critical angle for balance, and to increase the walker's moment of inertia.

   a. Use the diagrams below to describe how the critical angle is changed by the tightrope walker's use of the pole.

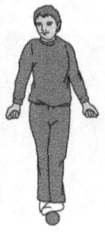   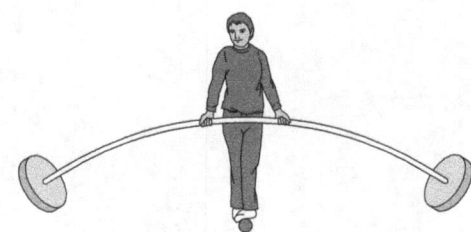

   b. Why would increasing the tightrope walker's moment of inertia also help to make him less likely to fall?

8.3 Springs and Hooke's Law

10. The graph below shows the stretching of two different springs, A and B, when different forces were applied.

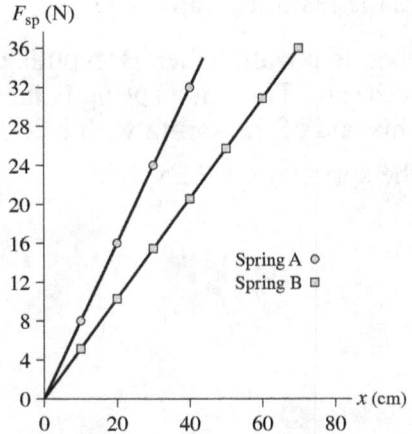

a. Which spring is stiffer? That is, which spring requires a larger pull to get the same amount of stretch? Explain how you can tell by looking at the graph.

b. Determine the spring constant for each spring.

$k_A =$ _____

$k_B =$ _____

11. A spring is attached to the floor and pulled straight up by a string. The spring's tension is measured. The graph shows the tension in the spring as a function of the spring's length L.

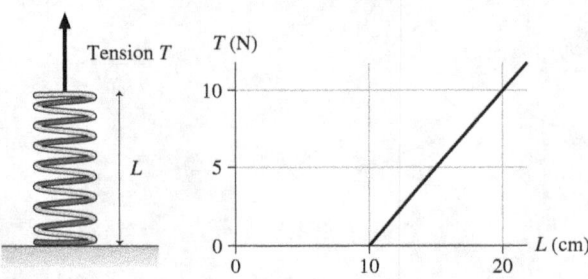

a. Does this spring obey Hooke's Law? Explain why or why not.

b. If it does, what is the spring constant?

12. A spring has an unstretched length of 10 cm. It exerts a restoring force F when stretched to a length of 11 cm.

    a. For what length of the spring is its restoring force $3F$? _____

    b. At what compressed length is the restoring force $2F$? _____

13. The left end of a spring is attached to a wall. When Bob pulls on the right end with a 200 N force, he stretches the spring by 20 cm. The same spring is then used for a tug-of-war between Bob and Carlos. Each pulls on his end of the spring with a 200 N force.

    a. How far does Bob's end of the spring move? Explain.

    b. How far does Carlos's end of the spring move? Explain.

14. A weight hung from a spring stretches the spring by 4.0 cm. If the same weight is hung from a second spring having half the spring constant as the first, by how much will the second spring stretch? Explain.

8.4 Stretching and Compressing Materials

8.5 Forces and Torques in Bodies

15. A force stretches a wire by 1 mm.
    a. A second wire of the same material has the same cross section and twice the length. How far will it be stretched by the same force? Explain.

    b. A third wire of the same material has the same length and twice the diameter as the first. How far will it be stretched by the same force? Explain.

16. A 2000 N force stretches a wire by 1 mm.
    a. A second wire of the same material is twice as long and has twice the diameter. How much force is needed to stretch it by 1 mm? Explain.

    b. A third wire is twice as long as the first and has the same diameter. How far is it stretched by a 4000 N force?

17. A wire is stretched right to the breaking point by a 5000 N force. A longer wire made of the same material has the same diameter. Is the force that will stretch it right to the breaking point larger than, smaller than, or equal to 5000 N? Explain.

You Write the Problem!

Exercises 18–19: You are given the equation that is used to solve a problem. For each of these:

a. Write a *realistic* physics problem for which this is the correct equation. Look at worked examples and end-of-chapter problems in the textbook to see what realistic physics problems are like. Be sure that the problem you write, and the answer you ask for, is consistent with the information given in the equation.
b. Draw a free-body diagram or force diagram for your problem.
c. Finish the solution of the problem.

18. $n + T - 80\,\text{N} = 0$
$(0.60\,\text{m})\,T - (0.50\,\text{m})(80\,\text{N}) = 0$

19. $k(0.050\,\text{m}) - (0.15)(0.12\,\text{kg})(9.8\,\text{m/s}^2) = (0.12\,\text{kg})(0.65\,\text{m/s}^2)$

9 Momentum

9.1 Impulse

9.2 Momentum and the Impulse-Momentum Theorem

1. What impulse is delivered by each of these forces?

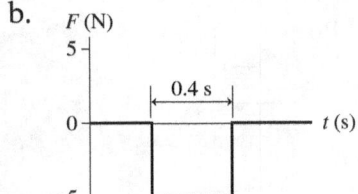

a.

b.

c.

2. Rank in order, from largest to smallest, the momenta $(p_x)_1$ to $(p_x)_5$.

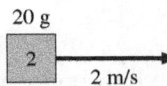 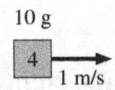 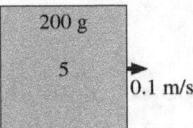

Order:

Explanation:

3. The position-versus-time graph is shown for a 500 g object. Draw the corresponding momentum-versus-time graph. Include an appropriate vertical scale.

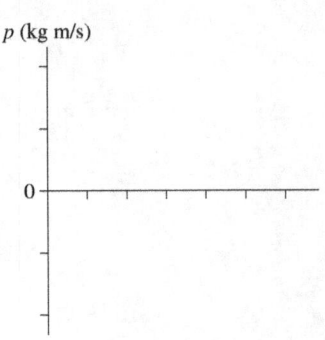

4. The momentum-versus-time graph is shown for a 500 g object. Draw the corresponding acceleration-versus-time graph. Include an appropriate vertical scale.

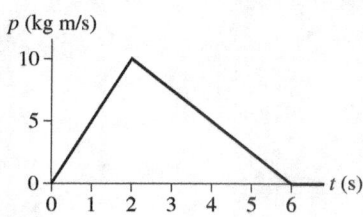 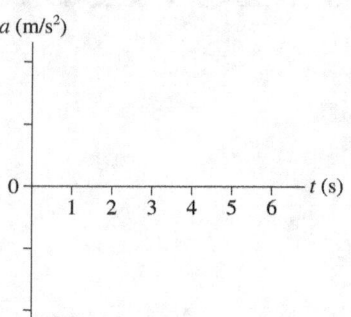

5. In each of the following, where a rubber ball bounces with no loss of speed, is the change in momentum Δp positive (+), negative (−), or zero (0)? Explain.

a.

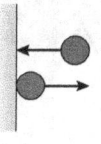

$\Delta p_x =$ _____

b.

$\Delta p_x =$ _____

c.

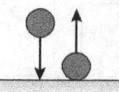

$\Delta p_y =$ _____

6. A 2 kg object is moving to the right with a speed of 1 m/s when it experiences an impulse due to the force shown in the graph. What is the object's speed and direction after the impulse?

a.

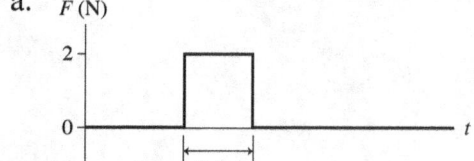

b.

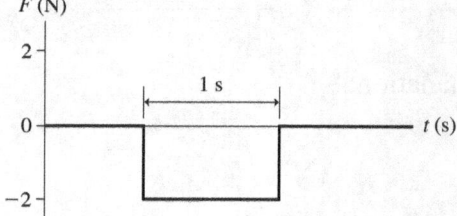

7. A carnival game requires you to knock over a wood post by throwing a ball at it. You're offered a very bouncy rubber ball and a very sticky clay ball of equal mass. Assume that you can throw them with equal speed and equal accuracy. You only get one throw.

   a. Which ball will you choose? Why?

   b. Let's think about the situation more carefully. Both balls have the same initial momentum $(p_x)_i$ just before hitting the post. The clay ball sticks, the rubber ball bounces off with essentially no loss of speed. What is the final momentum of each ball?

   Clay ball: $(p_x)_f =$ _____   Rubber ball: $(p_x)_f =$ _____
   Hint: Momentum has a sign. Did you take the sign into account?

   c. What is the *change* in the momentum of each ball?

   Clay ball: $\Delta p_x =$ _____   Rubber ball: $\Delta p_x =$ _____

   d. Which ball experiences a larger impulse during the collision? Explain.

   e. From Newton's third law, the impulse that the ball exerts on the post is equal in magnitude, although opposite in direction, to the impulse that the post exerts on the ball. Which ball exerts the larger impulse on the post?

   f. Don't change your answer to part a, but are you still happy with that answer? If not, how would you change your answer? Why?

8. A small, light ball S and a large, heavy ball L move toward each other, collide, and bounce apart.

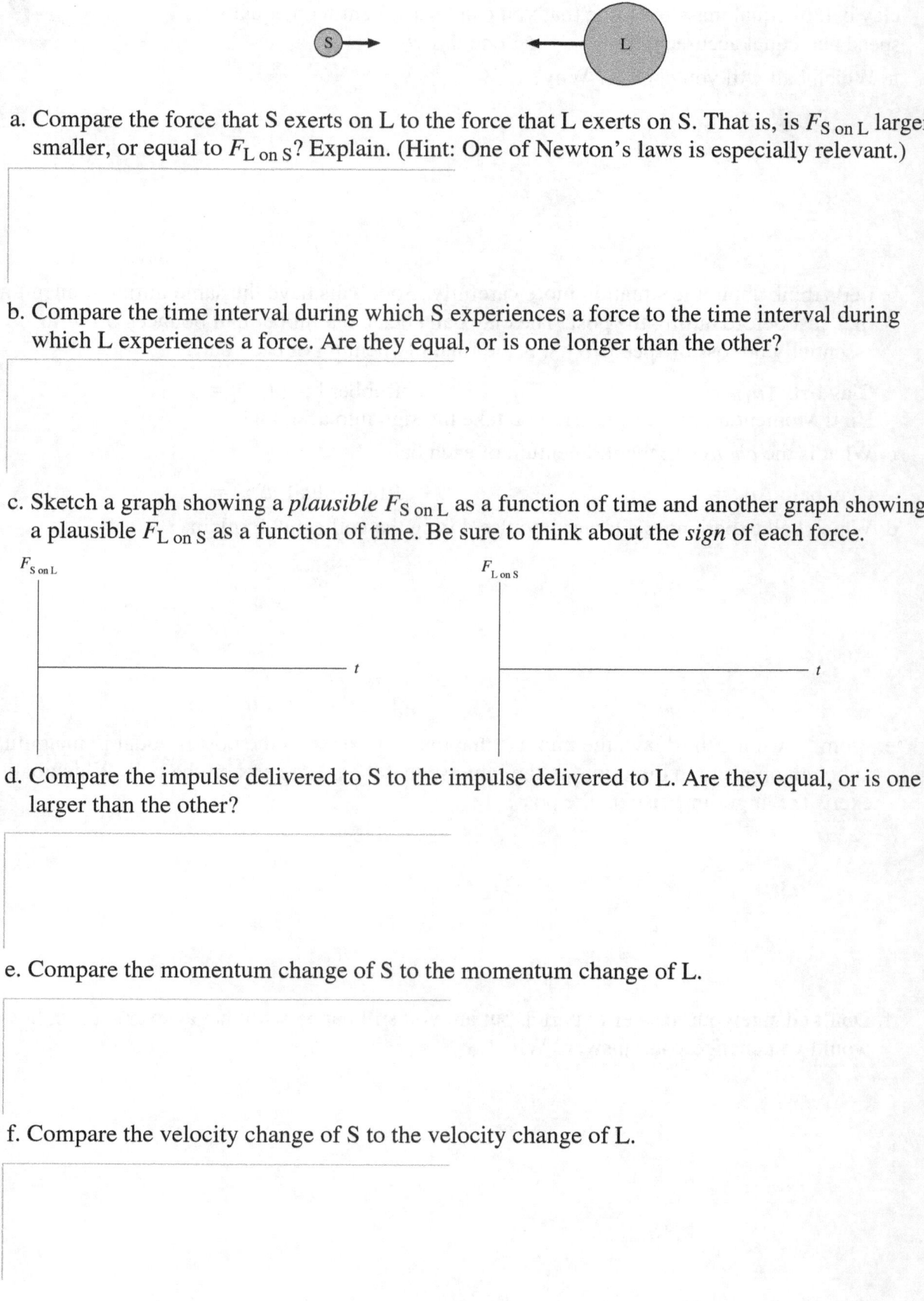

a. Compare the force that S exerts on L to the force that L exerts on S. That is, is $F_{\text{S on L}}$ larger, smaller, or equal to $F_{\text{L on S}}$? Explain. (Hint: One of Newton's laws is especially relevant.)

b. Compare the time interval during which S experiences a force to the time interval during which L experiences a force. Are they equal, or is one longer than the other?

c. Sketch a graph showing a *plausible* $F_{\text{S on L}}$ as a function of time and another graph showing a plausible $F_{\text{L on S}}$ as a function of time. Be sure to think about the *sign* of each force.

$F_{\text{S on L}}$ t $F_{\text{L on S}}$ t

d. Compare the impulse delivered to S to the impulse delivered to L. Are they equal, or is one larger than the other?

e. Compare the momentum change of S to the momentum change of L.

f. Compare the velocity change of S to the velocity change of L.

9.3 Solving Impulse and Momentum Problems

Exercises 9–11: Prepare a before-and-after visual overview for these problems, but *do not* solve them.

- Draw pictures of "before" and "after."
- Establish a coordinate system.
- Define symbols relevant to the problem.
- List known information, *and* identify the desired unknown.

9. A 50 kg archer, standing on frictionless ice, shoots a 100 g arrow at a speed of 100 m/s. What is the recoil speed of the archer?

10. The parking brake on a 2000 kg Cadillac has failed, and it is rolling slowly, at 1 mph, toward a group of small innocent children. As you see the situation, you realize there is just time for you to drive your 1000 kg Volkswagen head-on into the Cadillac and thus save the children. With what speed should you impact the Cadillac to bring it to a halt?

11. Dan is gliding on his skateboard at 4 m/s. He suddenly jumps backward off the skateboard, kicking the skateboard forward at 8 m/s. How fast is Dan going as his feet hit the ground? Dan's mass is 50 kg and the skateboard's mass is 5 kg.

12. Blocks A and B, both initially at rest, are pushed to the right continuously by identical constant forces. Block B is more massive than Block A. Which block crosses the finish line with more momentum? Or do they finish with equal momenta? Explain.

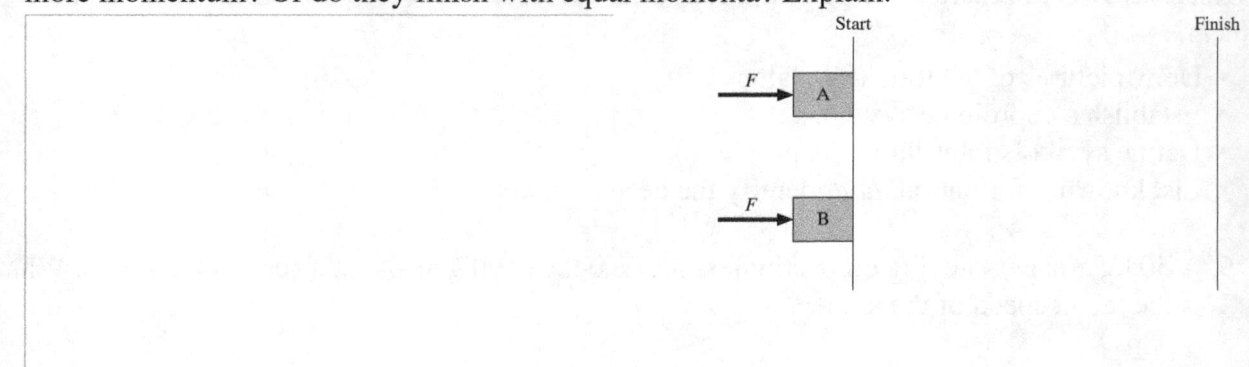

13. Blocks A and B, of equal mass, are pushed to the right continuously by identical constant forces. Block B starts from rest, but Block A is already moving to the right as it crosses the starting line. Which block undergoes a larger *change* in momentum before crossing the finish line? Or are they the same? Explain.

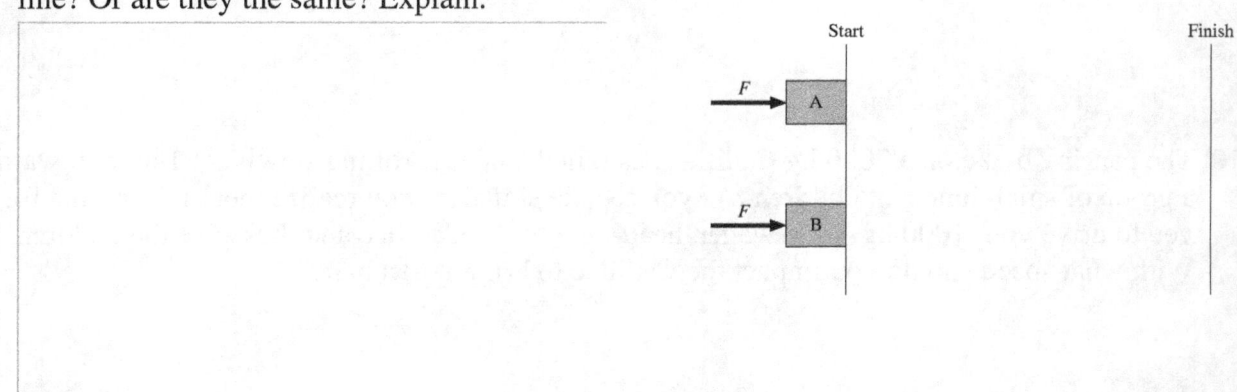

14. Blocks A and B are pushed to the right continuously by identical constant forces for exactly 1.0 s, starting from rest. Block B is more massive than Block A. After 1.0 s, which block has more momentum? Or do they have equal momenta? Explain.

9.4 Conservation of Momentum

15. As you release a ball, it falls—gaining speed and momentum. Is momentum conserved?

   a. Answer this question from the perspective of choosing the ball alone as the system.

   b. Answer this question from the perspective of choosing a ball + Earth as the system.

16. Two particles collide, one of which was initially moving and the other initially at rest.

   a. Is it possible for *both* particles to be at rest after the collision? Give an example in which this happens, or explain why it can't happen.

   b. Is it possible for *one* particle to be at rest after the collision? Give an example in which this happens, or explain why it can't happen.

17. A tennis ball traveling to the left at speed v_{Bi} is hit by a tennis racket moving to the right
    PSA at speed v_{Ri}. Although the racket is swung in a circular arc, its forward motion during the
    9.1 collision with the ball is so small that we can consider it to be moving in a straight line.
    Further, we can invoke the *impulse approximation* to neglect the steady force of the arm on the
    racket during the brief duration of its collision with the ball. Afterward, the ball is returned to
    the right at speed v_{Bf}. What is the racket's speed after it hits the ball? The masses of the ball
    and racket are m_B and m_R, respectively.

    a. Begin by drawing a before-and-after visual overview, as described in Tactics Box 9.1. You
       can assume that the racket continues in the forward direction but at a reduced speed.

    b. Define the system. That is, what object or objects should be inside the system so that it is an
       isolated system whose momentum is conserved?

    c. Write an expression for $(P_x)_i$, the total momentum of the system before the collision. Your
       expression should be written using the quantities given in the problem statement. Notice,
       however, that you're given *speeds*, but momentum is defined in terms of *velocities*. Based
       on your coordinate system and the directions of motion, you may need to give a negative
       momentum to one or more objects.

    d. Now write an expression for $(P_x)_f$, the total momentum of the system after the collision.

    e. If you chose the system correctly, its momentum is conserved. So equate your expressions
       for the initial and final total momentum, and then solve for what you want to find.

9.5 Inelastic Collisions

18. Determine whether each of the following graphs represents an inelastic collision between Object A (solid line) and Object B (dashed line). The objects in an inelastic collision must stick together, *and* the collision must conserve momentum. Part of your explanation should consider the relative masses of A and B.

Note that part a is a position-versus-time graph, but parts b–d are velocity-versus-time graphs.

a.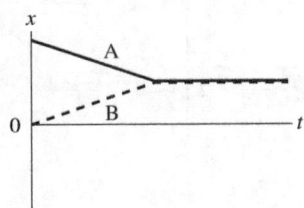

b.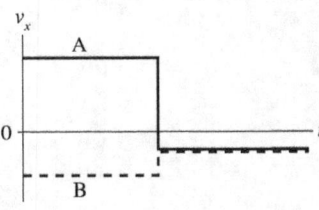

c.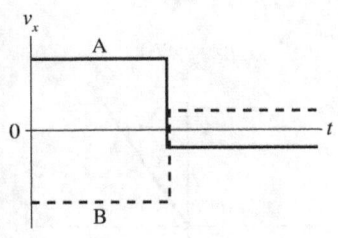

d.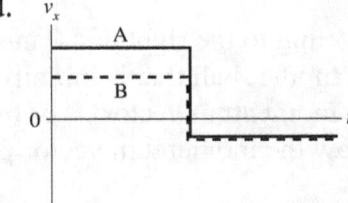

9.6 Momentum and Collisions in Two Dimensions

19. An object initially at rest explodes into three fragments. The momentum vectors of two of the fragments are shown. Draw the momentum vector $\vec{p}_3$ of the third fragment.

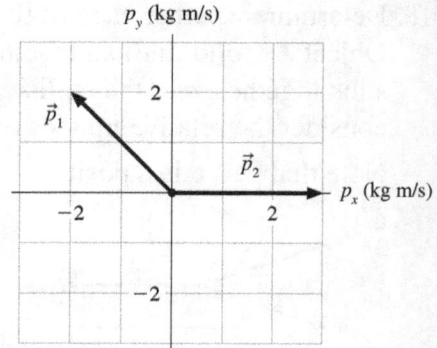

20. An object initially at rest explodes into three fragments. The momentum vectors of two of the fragments are shown. Draw the momentum vector $\vec{p}_3$ of the third fragment.

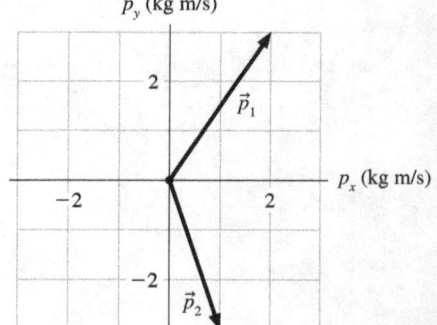

21. A 500 g ball traveling to the right at 4.0 m/s collides with and bounces off another ball that is initially at rest. The figure shows the momentum vector $\vec{p}_1$ of one ball after the collision. Draw the momentum vector $\vec{p}_2$ of the second ball.

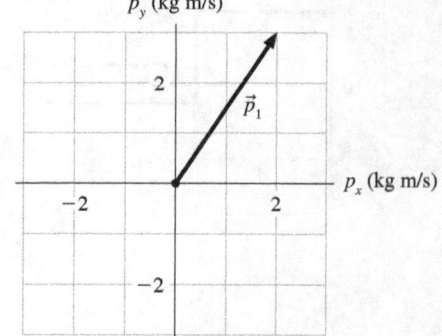

22. A 500 g ball traveling to the right at 4.0 m/s collides with and bounces off another ball that is initially at rest. The figure shows the momentum vector $\vec{p}_1$ of one ball after the collision. Draw the momentum vector $\vec{p}_2$ of the other ball.

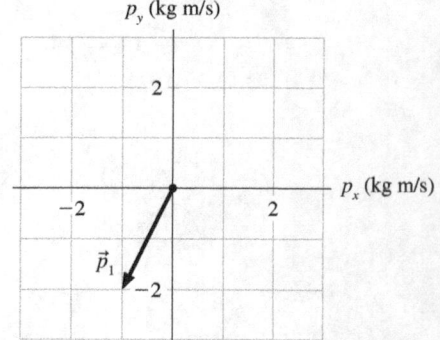

9.7 Angular Momentum

23. An isolated hoop of mass M and radius R is rotating with an angular speed of 60 rpm about its axis.

    a. What would be its angular speed if its mass suddenly doubled without changing its radius? Explain.

    b. What would be its angular speed if its radius suddenly doubled without changing its mass? Explain.

24. A solid circular disk and a circular hoop (like a bicycle wheel) both have mass M and radius R. If both are rotating with the same angular velocity ω, which has the larger angular momentum? Or are they equal? Explain.

You Write the Problem!

Exercises 25–28: You are given the equation that is used to solve a problem. For each of these:

  a. Write a *realistic* physics problem for which this is the correct equation. Look at worked examples and end-of-chapter problems in the textbook to see what realistic physics problems are like. Be sure that the problem you write, and the answer you ask for, is consistent with the information given in the equation.
  b. Draw a before-and-after visual overview for your problem.
  c. Finish the solution of the problem.

25. $(0.10 \text{ kg})(40 \text{ m/s}) - (0.10 \text{ kg})(-30 \text{ m/s}) = \frac{1}{2}(1400 \text{ N}) \Delta t$

26. $(600 \text{ g})(4.0 \text{ m/s}) = (400 \text{ g})(3.0 \text{ m/s}) + (200 \text{ g})(v_{2x})_i$

27. $(3000 \text{ kg})(v_x)_f = (2000 \text{ kg})(5.0 \text{ m/s}) + (1000 \text{ kg})(-4.0 \text{ m/s})$

28. $(50 \text{ g})(v_{1x})_f + (100 \text{ g})(7.5 \text{ m/s}) = (150 \text{ g})(1.0 \text{ m/s})$

10 Energy and Work

10.1 The Basic Energy Model

1. What are the two primary processes by which energy can be transferred from the environment to a system?

2. Identify the energy *transformations* in each of the following processes (e.g., $K \rightarrow U_g \rightarrow E_{th}$)

   a. A ball is dropped from atop a tall building.

   b. A helicopter rises from the ground at constant speed.

   c. An arrow is shot from a bow and stops in the center of its target.

   d. A pole vaulter runs, plants his pole, and vaults up over the bar.

3. Identify the energy *transfers* that occur in each of the following processes (e.g., $W \rightarrow K$).

   a. You pick up a book from the floor and place it on a table.

   b. You roll a bowling ball.

   c. You sand a board with sandpaper, making the board and the sandpaper warm.

10.2 Work

4. For each situation described below:
   - Draw a before-and-after diagram, similar to Figures 10.8 and 10.11 in the textbook.
   - Show and identify *all* forces acting on the object.
   - Fill in the table by listing each of your identified forces, by name or symbol, and showing whether the sign of the work done by that force is positive (+), negative (−), or zero (0).

a. An elevator suspended by a single cable moves upward.

Force	Sign of W
___	___
___	___
___	___
___	___

b. An elevator suspended by a single cable moves downward.

Force	Sign of W
___	___
___	___
___	___
___	___

c. A rope pulls a box to the left across a frictionless floor.

Force	Sign of W
___	___
___	___
___	___
___	___

5. An object experiences a force while undergoing the displacement shown. Is the work done positive (+), negative (−), or zero (0)?

a.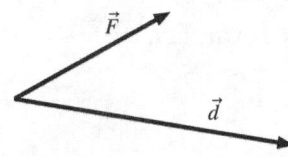

Sign of $W = $ _____

b.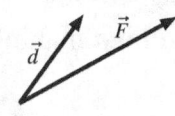

Sign of $W = $ _____

c.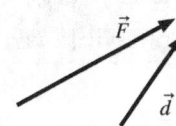

Sign of $W = $ _____

d.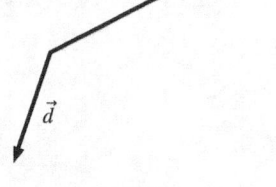

Sign of $W = $ _____

e.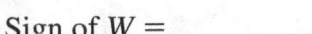

Sign of $W = $ _____

f.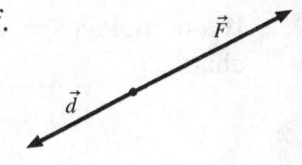

Sign of $W = $ _____

6. Each of the diagrams below shows a displacement vector for an object. On the figure, draw and label a force vector that will do work on the object with the sign indicated.

a. $W > 0$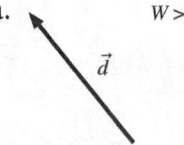

b. $W < 0$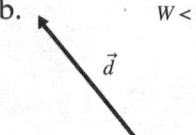

c. $W = 0$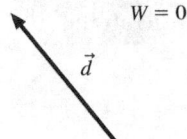

7. Equal forces push together the two equal-mass boxes shown in the figure. The table surface is frictionless. Both boxes have the same initial speed, and later both have the same slower speed. Let the system consist of the boxes and the spring.

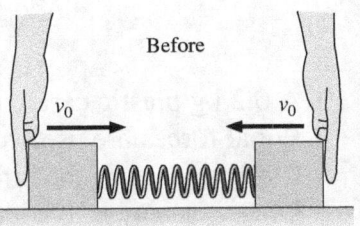

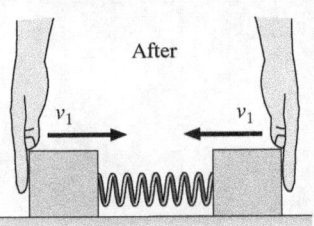

a. Is the work done on the system positive, negative, or zero? Explain.

b. Is ΔE, the change in the system's energy, positive, negative, or zero? Explain.

10.3 Kinetic Energy

8. Can kinetic energy ever be negative? _____

Give a plausible *reason* for your answer without making use of any formulas.

9. a. If a particle's speed increases by a factor of three, by what factor does its kinetic energy change?

   b. Particle A has half the mass and eight times the kinetic energy of particle B. What is the speed ratio v_A/v_B?

10. A 0.2 kg plastic cart and a 20 kg lead cart both roll without friction on a horizontal surface. Equal forces are used to push both carts forward a distance of 1 m, starting from rest. After traveling 1 m, is the kinetic energy of the plastic cart greater than, less than, or equal to the kinetic energy of the lead cart? Explain.

10.4 Potential Energy

11. Below we see a 1 kg object that is initially 1 m above the ground and rises to a height of 2 m. Anjay and Brittany each measure its position but use a different coordinate system to do so. Fill in the table to show the initial and final gravitational potential energies and ΔU_g as measured by Anjay and Brittany.

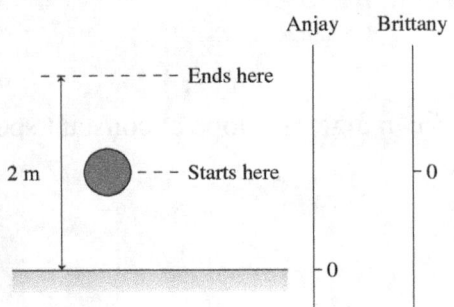

	$(U_g)_i$	$(U_g)_f$	ΔU_g
Anjay			
Brittany			

12. Rank in order, from most to least, the amount of elastic potential energy $(U_s)_1$ to $(U_s)_4$ stored in each of these springs.

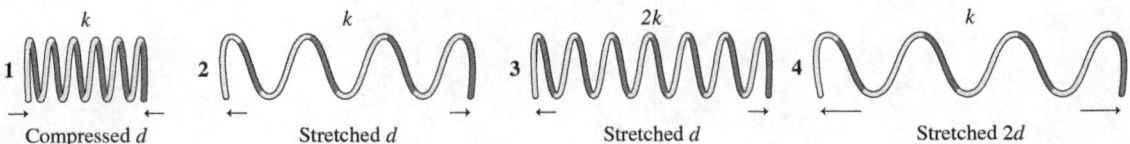

Order:

Explanation:

13. A heavy object is released from rest at position 1 above a spring. It falls and contacts the spring at position 2. The spring achieves maximum compression at position 3. Fill in the table below to indicate whether each of the quantities are +, −, or 0 during the intervals 1→2, 2→3, and 1→3.

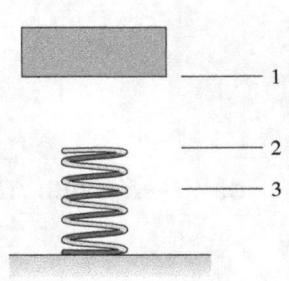

	1→2	2→3	1→3
ΔK			
ΔU_g			
ΔU_s			

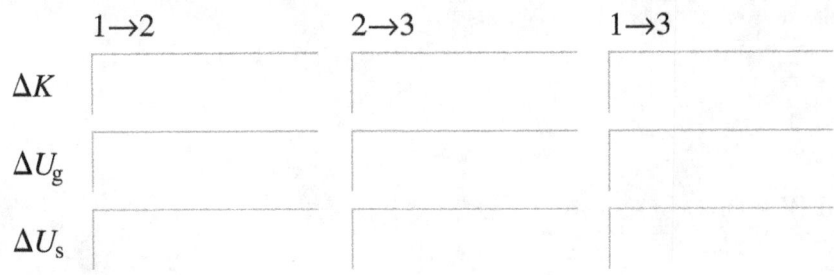

10.5 Thermal Energy

14. A car traveling at 60 mph slams on its brakes and skids to a halt. What happened to the kinetic energy the car had just before stopping?

15. What energy transformations occur as a skier glides down a gentle slope at constant speed?

16. Give a *specific* example of a situation in which:
    a. $W \rightarrow K$ with $\Delta U = 0$ and $\Delta E_{th} = 0$.

    b. $W \rightarrow U$ with $\Delta K = 0$ and $\Delta E_{th} = 0$.

    c. $K \rightarrow U$ with $W = 0$ and $\Delta E_{th} = 0$.

    d. $W \rightarrow E_{th}$ with $\Delta K = 0$ and $\Delta U = 0$.

    e. $U \rightarrow E_{th}$ with $\Delta K = 0$ and $W = 0$.

10.6 Conservation of Energy

17. What is meant by an *isolated system*?

18. Identify an appropriate system for applying conservation of energy to each of the following:

    a. A compressed spring is used to launch a ball into the air.

    System:

    b. A compressed spring is used to push a cart on a frictionless air track.

    System:

    c. A compressed spring is used to push a block across a table, where it then slows and stops.

    System:

19. Below are shown three frictionless tracks. A block is released from rest at the position shown on the left. To which point does the block make it on the right before reversing direction and sliding back? Point B is the same height as the starting position.

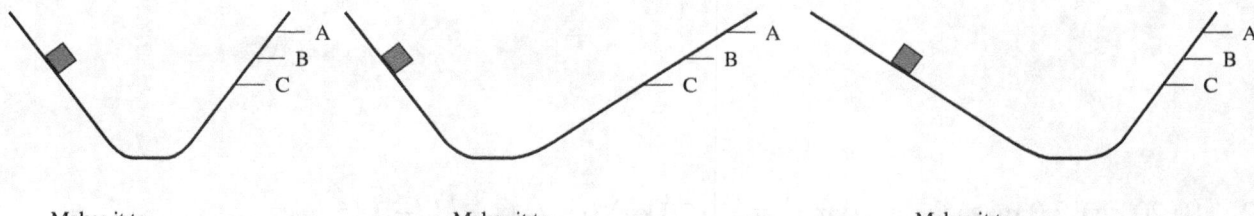

Makes it to _____   Makes it to _____   Makes it to _____

20. A spring gun shoots out a plastic ball at speed v_0. The spring is then compressed twice the distance it was on the first shot. By what factor is the ball's launch speed increased? Explain.

You Write the Problem!

Exercises 21–22: You are given the equation that is used to solve a problem. For each of these:

a. Write a *realistic* physics problem for which this is the correct equation. Look at worked examples and end-of-chapter problems in the textbook to see what realistic physics problems are like. Be sure that the problem you write, and the answer you ask for, is consistent with the information given in the equation.

b. Draw a before-and-after visual overview for your problem.

c. Finish the solution of the problem.

21. $\frac{1}{2}(1.5\,\text{kg})(5.0\,\text{m/s})^2 + (1.5\,\text{kg})(9.8\,\text{m/s}^2)(10\,\text{m})$
$= \frac{1}{2}(1.5\,\text{kg})(v_i)^2 + (1.5\,\text{kg})(9.8\,\text{m/s}^2)(0\,\text{m})$

22. $\frac{1}{2}(0.20\,\text{kg})(2.0\,\text{m/s})^2 + \frac{1}{2}k(0\,\text{m})^2 = \frac{1}{2}(0.20\,\text{kg})(0\,\text{m/s})^2 + \frac{1}{2}k(-0.15\,\text{m})^2$

23. A small cube of mass m slides back and forth in a frictionless,
PSA hemispherical bowl of radius R. Suppose the cube is released at
10.1 angle θ. What is the cube's speed at the bottom of the bowl?

a. Begin by drawing a before-and-after visual overview. Let the
cube's initial position and speed be y_i and v_i. Use a similar
notation for the final position and speed.

b. At the initial position, is either K_i or $(U_g)_i$ zero? If so, which? _____

c. At the final position, is either K_f or $(U_g)_f$ zero? If so, which? _____

d. Does thermal energy need to be considered in this situation? Why or why not?

e. Write the conservation of energy equation in terms of position and speed variables, omitting
any terms that are zero.

f. You're given not the initial position but the initial angle. Do the geometry and trigonometry
to find y_i in terms of R and θ.

g. Use your result of part f in the energy conservation equation, and then finish solving the
problem.

10.7 Energy Diagrams

10.8 Molecular Bonds and Chemical Energy

24. The figure shows a potential-energy curve. Suppose a particle with energy E_1 is at position A and moving to the right.

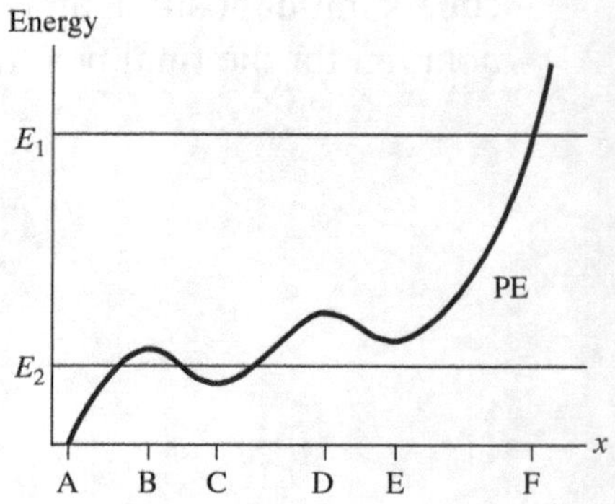

    a. For each of the following regions of the x-axis, does the particle speed up, slow down, maintain a steady speed, or change direction?

    A to B _____

    B to C _____

    C to D _____

    D to E _____

    E to F _____

    b. Where is the particle's turning point? _____

    c. For a particle that has energy E_2 what are the possible motions and where do they occur along the x-axis?

    d. What position or positions are points of stable equilibrium? For each, would a particle in equilibrium at that point have energy $\leq E_2$, between E_2 and E_1, or $\geq E_1$?

    e. What position or positions are points of unstable equilibrium? For each, would a particle in equilibrium at that point have energy $\leq E_2$, between E_2 and E_1, or $\geq E_1$?

10.9 Energy in Collisions

25. Ball 1 with an initial speed of 14 m/s has a perfectly elastic collision with Ball 2 that is initially at rest. Afterward, the speed of Ball 2 is 21 m/s.

    a. What will be the speed of Ball 2 if the initial speed of Ball 1 is doubled?

    b. What will be the speed of Ball 2 if the mass of Ball 1 is doubled?

26. Consider a perfectly elastic collision in which a moving ball 1 strikes an initially stationary Ball 2.

    a. Under what circumstances, if any, will Ball 1 come to a stop?

    b. Under what circumstances, if any, will Ball 1 recoil backwards?

    c. Under what circumstances, if any, will Ball 1 continue moving forward?

27. You can dive into a swimming pool of water from a high diving board without being hurt, but to dive into an empty pool from a much lower distance might be fatal. Why the difference?

10.10 Power

28. a. If you push an object 10 m with a 10 N force in the direction of motion, how much work do you do on it?

   b. How much power must you provide to push the object in 1 s? In 10 s? In 0.1 s?

29. a. To push an object *twice as fast* with the same force, must your power output increase? If so, by what factor? Explain.

   b. To push an object *twice as far* with the same force and at the same speed, must your power output increase? If so, by what factor? Explain.

11 Using Energy

11.1 Transforming Energy

1. A device uses 500 J of chemical energy to generate 100 J of electric energy.
   a. What is the efficiency of this device?

   b. A second device is twice as efficient. How much electric energy can it generate from the same 500 J of chemical energy?

2. A small battery-operated car zips across the floor. In thinking about the efficiency of the car:
   a. What do you get? In particular, what useful form of energy increases?

   b. What do you pay? In particular, what form of energy decreases?

3. Is it possible for the efficiency of some device or machine to be greater than 1? Either give an example where the efficiency is greater than 1 or explain why it's not possible.

4. a. Given that efficiency is defined as e = (what you get)/(what you had to pay), how might one define the concept of "inefficiency"?

   b. Write a mathematical relationship to relate your definition of inefficiency, which you can denote as i, to the definition of efficiency e.

11.2 Energy in the Body

5. Food contains energy that your body can use. Is the energy in food kinetic, potential, thermal, chemical, or nuclear? Explain.

6. Can you run further by burning an ounce of fat or an ounce of carbohydrates? Explain.

7. The rate at which energy is used by a human while running is approximately proportional to the speed. If we expressed the energy used by a runner in terms of, say, miles per candy bar, who gets better mileage per candy bar, a runner who completes a marathon in $2\frac{1}{2}$ hours or one who completes the same race in 4 hours? Or is it the same for both runners? Explain.

8. Suppose your body burns 10 Cal climbing stairs. Which of the following will allow you to burn 20 Cal?
   a. Climb twice as high at the same speed.
   b. Climb the original stairs twice as fast.
   c. Either a or b.
   d. Neither a nor b.

   Explain.

11.3 Temperature, Thermal Energy, and Heat

9. Rank in order, from highest to lowest, the temperatures $T_1 = 0$ K, $T_2 = 0°C$, and $T_3 = 0°F$.

10. "Room temperature" is often considered to be 68°F. What is room temperature in °C and in K?

11. a. What is the average kinetic energy of atoms at absolute zero? _____

    b. Can an atom have negative kinetic energy? _____

    c. Is it possible to have a temperature less than absolute zero? Explain.

12. Do each of the following describe a property of a system, an interaction of a system with its environment, or both? Explain.

    a. Temperature:

    b. Heat:

    c. Thermal energy:

11.4 The First Law of Thermodynamics

13. For each of the following processes:

    a. Is the value of the work W, the heat Q, and the change of thermal energy ΔE_{th} positive (+), negative (−), or zero (0)?

    b. Does the temperature increase (+), decrease (−), or not change (0)?

    Make sure your answers are consistent with the first law of thermodynamics: $\Delta E_{th} = W + Q$.

	W	Q	ΔE_{th}	ΔT
• You hit a nail with a hammer. The system is the nail.				
• You hold a nail over a Bunsen. The system is the nail.				
• Expanding high-pressure steam spins a turbine. The system is the steam.				
• Steam contacts a cold surface and condenses. The system is the steam.				
• A piston rapidly compresses the gas in a cylinder. The system is the gas.				

14. Metal blocks A and B, both at room temperature, are balanced on a see-saw. Block B is removed, heated to just below its melting point, then returned to the exact same position on the see-saw. Afterward, is block A higher than, lower than, or still level with block B? Explain.

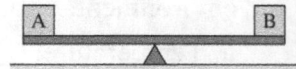

11.5 Heat Engines

15. Rank in order, from largest to smallest, the efficiencies e_1 to e_4 of these heat engines.

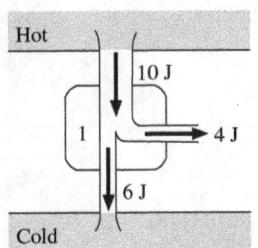 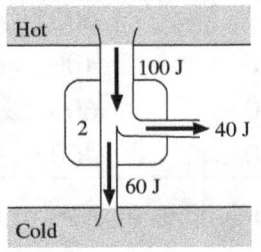 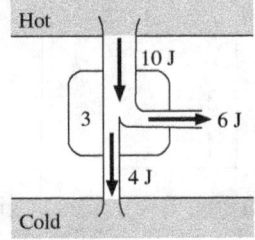 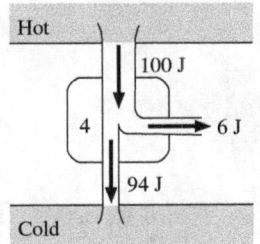

Order:

Explanation:

16. For each engine shown,
    a. Supply the missing value.
    b. Determine the efficiency.

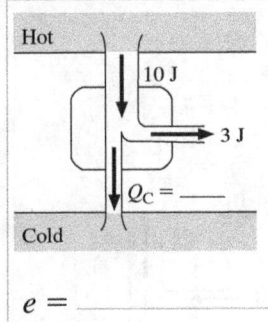 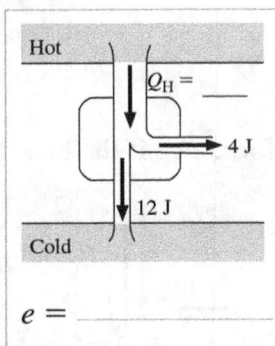 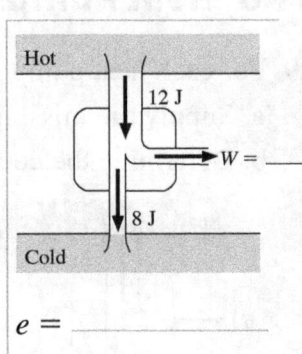

$e =$ _____ $e =$ _____ $e =$ _____

17. Efficiency is a dimensionless quantity, so why is it necessary to measure temperatures in Kelvin rather than °C or °F to determine efficiency when using the equation $e_{max} = 1 - \dfrac{T_C}{T_H}$?

18. Four heat engines with maximum possible efficiency operate with the hot and cold reservoir temperatures shown in the table.

Engine	T_C (K)	T_H (K)
1	300	600
2	200	400
3	200	600
4	300	400

Rank in order, from largest to smallest, the efficiencies e_1 to e_4 of these engines.

Order:

Explanation:

11.6 Heat Pumps

19. For each heat pump shown,
    a. Supply the missing value.
    b. Determine the coefficient of performance if the heat pump is used for cooling.

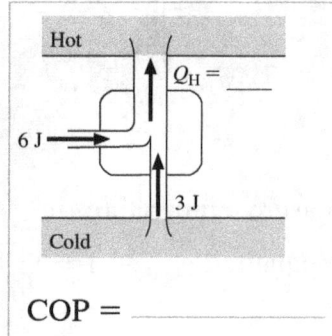

COP = _____

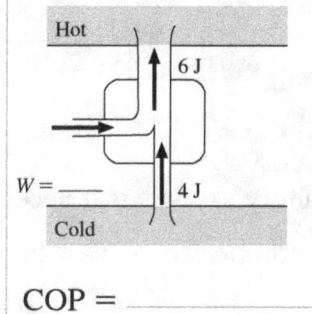

COP = _____

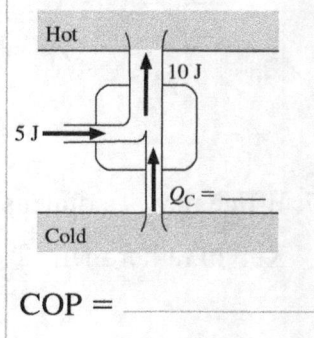

COP = _____

20. Does a refrigerator do work in order to cool the interior? Explain.

11.7 Entropy and the Second Law of Thermodynamics

11.8 Systems, Energy, and Entropy

21. If you place a jar of perfume in the center of a room and remove the stopper, you will soon be able to smell the perfume throughout the room. If you wait long enough, will all the perfume molecules ever be back in the jar at the same time? Why or why not?

22. Every cubic meter of air contains $\approx 200{,}000$ J of thermal energy. This is approximately the kinetic energy of a car going 40 mph. Even though it might be difficult to do, could a clever engineer design a car that uses the thermal energy already in the air as "fuel"? Even if only 1% of the thermal energy could be "extracted" from the air, it would take only ≈ 100 m^3 of air—the volume of a typical living room in a house—to get the car up to speed. Is this idea possible? Or does it violate the laws of physics?

23. Suppose you place an ice cube in a cup of room-temperature water and then seal them in a well-insulated container. No energy can enter or leave the container.

    a. If you open the container an hour later, which do you expect to find: a cup of water, slightly cooler than room temperature, or a large ice cube and some 100°C steam?

    b. Finding a large ice cube and some 100°C steam would not violate the first law of thermodynamics. $W = 0$ J and $Q = 0$ J, because the container is sealed, and $\Delta E_{th} = 0$ J because the increase in thermal energy of the water molecules that have become steam is offset by the decrease in water molecules that have turned to ice. Energy is conserved, yet we never see a process like this. Why not?

24. Are each of the following processes reversible or irreversible? Would the second law of thermodynamics be violated by any of the processes? Explain.

    a. A freshly baked pie cools on a window sill.

    b. A neatly raked pile of leaves is scattered by the wind.

    c. The wind gathers up the fallen leaves in a yard and leaves them in a neat pile.

12 Thermal Properties of Matter

12.1 The Atomic Model of Matter

1. The block shown is made up of eight identical cubes, set $2 \times 2 \times 2$.

   a. In the space at right, draw a block that has twice the total volume but still made out of cubes of the same size.

   b. How many cubes are needed to construct the larger block? _____
   c. How many cubes are needed to construct a solid block that is twice as large in each dimension ($4 \times 4 \times 4$)? _____
   d. By what factor is the volume of the block in part c greater than the volume of the original $2 \times 2 \times 2$ block? _____

2. a. Solids and liquids resist being compressed. They are not totally incompressible, but it takes large forces to compress them even slightly. If it is true that matter consists of atoms, what can you infer about the microscopic nature of solids and liquids from their incompressibility?

   b. Solids also resist being pulled apart. You can break a metal or glass rod by pulling the ends in opposite directions, but it takes a large force to do so. What can you infer from this observation about the properties of atoms?

12.2 The Atomic Model of an Ideal Gas

3. Gases, in contrast with solids and liquids, are very compressible. What can you infer from this observation about the microscopic nature of gases?

4. If you double the temperature of a gas,
   a. Does the root-mean-square speed of the atoms increase by a factor of $(2)^{1/2}$, 2, or 2^2? Explain.

   b. Does the average kinetic energy of the atoms increase by a factor of $(2)^{1/2}$, 2, or 2^2? Explain.

5. Suppose you could suddenly increase the speed of every atom in a gas by a factor of 2.
   a. Would the rms speed of the atoms increase by a factor of $(2)^{1/2}$, 2, or 2^2? Explain.

   b. Would the thermal energy of the gas increase by a factor of $(2)^{1/2}$, 2, or 2^2? Explain.

   c. Would the temperature of the gas increase by a factor of $(2)^{1/2}$, 2, or 2^2? Explain.

6. Lithium vapor, which is produced by heating lithium to a relatively low boiling point of 1340°C, forms a gas of Li_2 molecules. Each molecule has a molecular mass of 14 u. The molecules in nitrogen gas (N_2) have a molecular mass of 28 u. If the Li_2 and N_2 gases are at the same temperature, which of the following is true? (Circle the letter.)

   a. v_{rms} of N_2 = 2.00 × v_{rms} of Li_2.

   b. v_{rms} of N_2 = 1.41 × v_{rms} of Li_2.

   c. v_{rms} of N_2 = v_{rms} of Li_2.

   d. v_{rms} of N_2 = 0.71 × v_{rms} of Li_2.

   e. v_{rms} of N_2 = 0.50 × v_{rms} of Li_2.

   Explain.

7. The quantity y is proportional to the square root of x, and $y = 12$ when $x = 16$.

   a. Find y if $x = 9$: _____   b. Find x if $y = 6$: _____

   c. By what factor must x change for the value of y to double? _____

   d. Consider the equation in your text relating v_{rms} and T for an ideal gas. Which of these quantities plays the role of x in a square-root relationship $y = A\sqrt{x}$? Which plays the role of y?

    x is played by _____ y is played by _____

8. The quantity y is proportional to the square root of x. Initially $y = 10$. What is the value of y if the value of x is (a) doubled and (b) quadrupled?

9. A gas is held in a sealed container from which no molecules can enter or leave. Suppose the absolute temperature T of the gas is doubled. Will the gas pressure double? Why or why not can you draw this conclusion?

12.3 Ideal-Gas Processes

10. Consider an ideal gas contained in a confined volume. How would the pressure of the gas change if

    a. the number of molecules of the gas were doubled, without changing the container or the temperature?

    b. the volume of the container was doubled, without changing the number of molecules or the temperature?

    c. the temperature (in K) of the gas was doubled, without changing the number of molecules or the volume of the container?

    d. the rms speed of the molecules was doubled, without changing the number of molecules or the volume of the container?

11. The graphs below use a dot to show the initial state of a gas. Draw a pV diagram showing the following processes:

a. A constant-volume process that doubles the pressure.

b. An isobaric process that doubles the temperature.

c. An isothermal process that halves the volume.

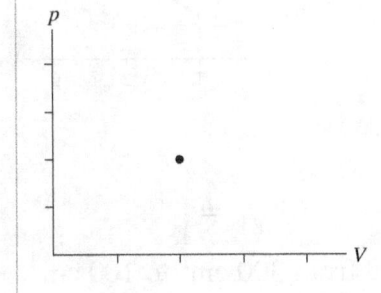 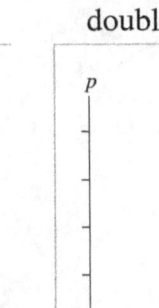 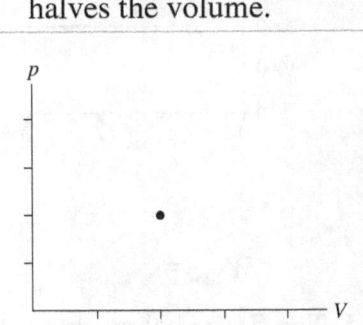

12. Interpret the pV diagrams shown below by
a. Naming the process.
b. Stating the *factors* by which p, V, and T change. (A fixed quantity changes by a factor of 1.)

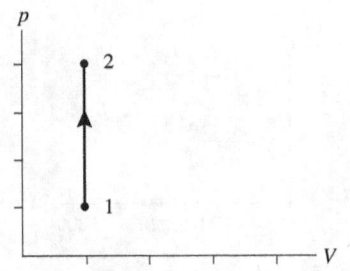 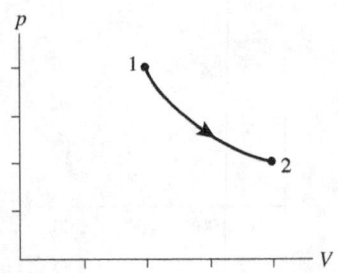 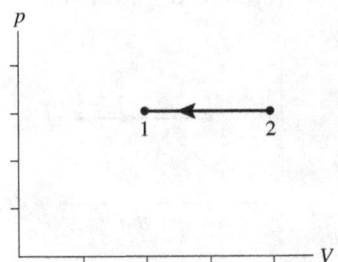

Process _____

p changes by _____

V changes by _____

T changes by _____

Process _____

p changes by _____

V changes by _____

T changes by _____

Process _____

p changes by _____

V changes by _____

T changes by _____

13. Starting from the initial state shown, draw a pV diagram for the three-step process:

i. A constant-volume process that halves the temperature, then

ii. An isothermal process that halves the pressure, then

iii. An isobaric process that doubles the volume.

Label each of the stages on your diagram.

14. How much work is done by the gas in each of the following processes?

a.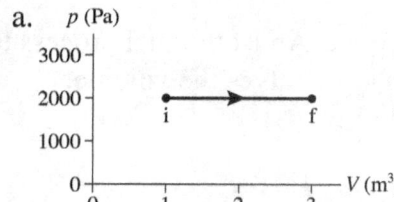

b.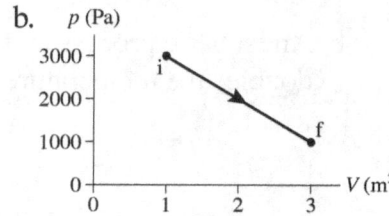

c.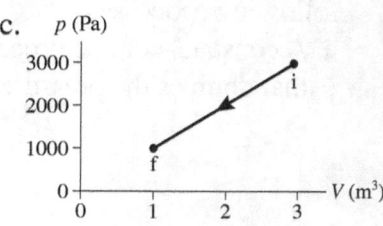

$W_{\text{gas}} =$ _____ $W_{\text{gas}} =$ _____ $W_{\text{gas}} =$ _____

15. The figure below shows a process in which a gas is compressed from 300 cm^3 to 100 cm^3.

   a. Use the middle set of axes to draw the pV diagram of a process that starts from initial state i, compresses the gas to 100 cm^3, and does the same amount of work on the gas as the process on the left.

   b. Is there a constant-volume process that does the same amount of work on the gas as the process on the left? If so, show it on the axes on the right. If not, use the blank space of the axes to explain why.

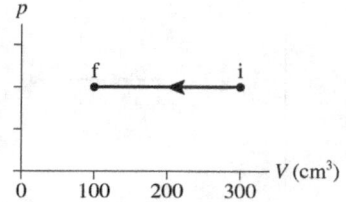 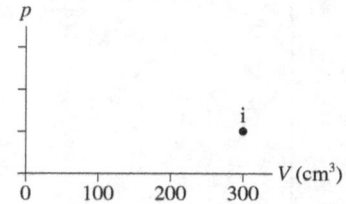 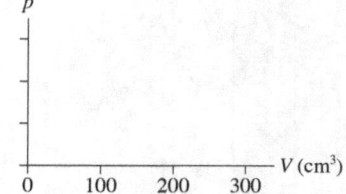

16. Starting from the point shown, draw a pV diagram for the following processes.

   a. An isobaric process in which work is done *by* the system.

   b. An adiabatic process in which work is done *on* the system.

   c. An isothermal process in which heat is *added to* the system.

   d. A constant-volume process in which heat is *removed from* the system.

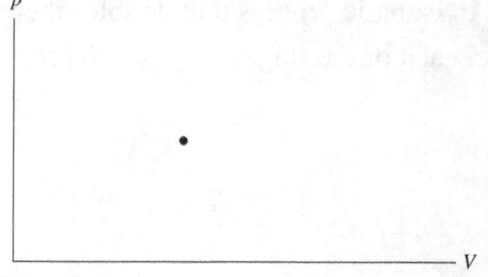

12.4 Thermal Expansion

17. The figure shows a thin square solid object with a circular hole in the center. The hole has a diameter that is half the length of each side of the square. Redraw the object in the space at right after the object has undergone an expansion of its size by 100% in each linear dimension.

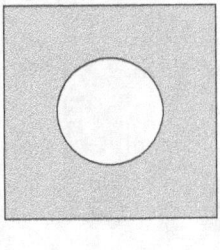

18. a. Solids are characterized by both a coefficient of linear expansion and a coefficient of volume expansion. Examine Table 12.3 and describe the relationship between these two coefficients.

    b. Liquids, in contrast with solids, have only a coefficient of volume expansion. Why don't liquids have a coefficient of linear expansion?

19. Containers A and B hold equal volumes of water at the same temperature. The temperatures of both containers are increased by the same amount. Does the height of the water in A *go up* by a larger amount, a smaller amount, or the same amount as the water goes up in B? Explain.

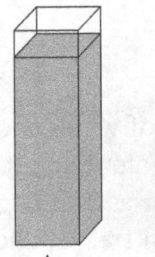 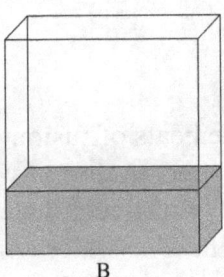

A   B

12.5 Specific Heat and Heat of Transformation

20. Explain in your own words the meaning of the statement that the specific heat of a liquid is 4000 J/kg·K.

21. Object A has a larger specific heat than object B. Which object's temperature will increase the most if both absorb the same amount of heat energy? Explain.

22. 100 g of water at 10°C and 100 g of water at 90°C each absorb the same amount of heat energy. Which sample's temperature increases the most? Or do they increase by the same amount? Explain. You can assume that the amount of heat absorbed is not enough to boil the 90°C sample.

23. The heats of fusion of water and nitrogen are 333 kJ/kg and 26 kJ/kg, respectively. The heats of vaporization of water and nitrogen are 2260 kJ/kg and 1990 kJ/kg, respectively. Suppose 100 g of liquid water at 0°C is poured into a flask of liquid nitrogen at its −196°C boiling point. The water will freeze, and in doing so it will boil off some nitrogen. Will the process of turning the 0°C water into solid ice at 0°C boil off less than 100 g, more than 100 g, or exactly 100 g of nitrogen? Explain.

12.6 Calorimetry

24. A beaker of water at 80.0°C is placed in the center of a large, well-insulated room whose air temperature is 20.0°C. Is the final temperature of the water:

    i.   20.0°C.
    ii. Slightly above 20.0°C.
    iii. 50.0°C.

    iv. Slightly less than 80.0°C.
    v.   80.0°C.

    Explain.

25. You have two 100 g cubes A and B, made of different materials. Cube A has a larger specific heat than cube B. Cube A, initially at 0°C, is placed in good thermal contact with cube B, initially at 200°C. The cubes are inside a well-insulated container where they don't interact with their surroundings. Is their final temperature greater than, less than, or equal to 100°C? Explain.

You Write the Problem!

Exercise 26: You are given the equation that is used to solve a problem.

   a. Write a *realistic* physics problem for which this is the correct equation. Look at worked examples and end-of-chapter problems in the textbook to see what realistic physics problems are like.
   b. Finish the solution of the problem.

26. $(2.72 \text{ kg})(140 \text{ J/kg} \cdot \text{K})(90°C - 15°C) + (0.50 \text{ kg})(449 \text{ J/kg} \cdot \text{K})(90°C - T_i) = 0$

12.7 Specific Heats of Gases

27. Explain *why* the molar specific heat at constant pressure is larger than the molar specific heat at constant volume.

28. Explain why the molar specific heat at constant volume of diatomic gases is larger than that of monatomic gases.

29. You need to raise the temperature of a gas by 10°C. To use the least amount of heat energy, should you heat the gas at constant pressure or at constant volume? Explain.

30. The pV diagram shows two processes, A and B, that take an ideal gas from initial state i to final state f.
    a. Is more work done by the gas in process A or in process B? Or is W_{gas} the same for both? Explain.

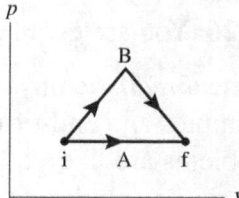

    b. According to the first law of thermodynamics, heat energy is required for both processes. Is more heat energy required for process A or for process B? Or is Q the same for both? Explain.

12.8 Heat Transfer

31. Heat flows through a cylindrical copper bar at the rate of 100 W. A second cylindrical copper bar between the same two temperatures is twice as long and has twice the diameter as the first bar. What is the rate of heat flow through the second bar?

32. The thermal conductivities of wood and concrete are 0.2 W/m·K and 0.8 W/m·K, respectively. Suppose your house has 1-inch-thick wood siding on the outside. You would like to replace the wood siding with a concrete exterior wall. What thickness must the concrete have to provide the same insulating value as the wood? Explain.

33. A titanium cube at 400 K radiates 50 W of heat. How much heat does the cube radiate if its temperature is increased to 800 K?

34. Sphere A has radius R and is at temperature T. Sphere B has radius $R/2$ and is at temperature $2T$. Which sphere radiates more heat? Explain.

12.9 Diffusion

35. Molecules from a very small source diffuse 1 cm in 1 s. How long will it take the molecules to diffuse 2 cm? Explain.

36. Molecules diffuse through a tube connecting two containers at the rate $2 \times 10^{16}\,\mathrm{s}^{-1}$. What will the diffusion rate be if
    a. The concentration difference is doubled?

    b. The length of the tube is doubled?

    c. The radius of the tube is doubled?

13 Fluids

13.1 Fluids and Density

1. An object has density ρ.

   a. Suppose each of the object's three dimensions is increased by a factor of 2 without changing the material of which the object is made. Will the density change? If so, by what factor? Explain.

   b. Suppose each of the object's three dimensions is increased by a factor of 2 without changing the object's mass. Will the density change? If so, by what factor? Explain.

2. A stone cutter cuts a section off of a marble slab that is one-eighth the weight of the original slab. Is the density of the cut section less than, more than, or the same as the density of the original slab? Explain.

3. A cylinder contains 2 g of oxygen gas. A piston is used to compress the gas. After the gas has been compressed:

   a. Has the mass of the gas increased, decreased, or not changed? Explain.

   b. Has the density of the gas increased, decreased, or not changed? Explain.

4. Air enclosed in a cylinder has density $\rho = 1.4 \text{ kg/m}^3$.

   a. What will be the density of the air if the length of the cylinder is doubled while the radius is unchanged?

   b. What will be the density of the air if the radius of the cylinder is halved while the length is unchanged?

13.2 Pressure

5. Rank in order, from largest to smallest, the pressures p_A to p_D in containers A through D at the depths indicated by the dashed line.

Order:

Explanation:

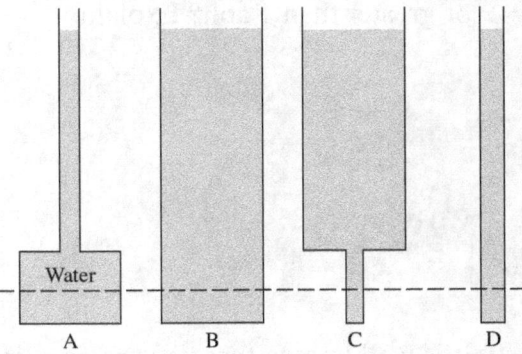

6. A and B are rectangular tanks full of water. They have equal depths, equal thicknesses (the dimension into the page) but different widths.

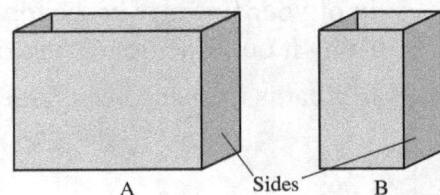

   a. Compare the forces the water exerts on the bottoms of the tanks. Is F_A larger than, smaller than, or equal to F_B? Explain.

   b. Compare the forces the water exerts on the sides of the tanks. Is F_A larger than, smaller than, or equal to F_B? Explain.

7. Is p_A larger than, smaller than, or equal to p_B? Explain.

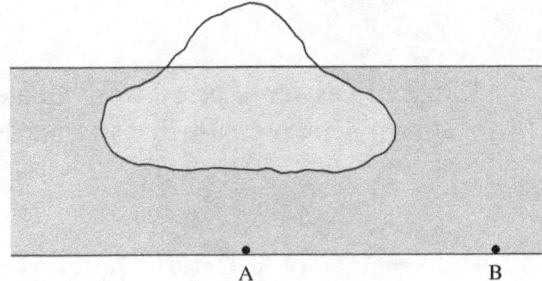

8. The figure shows a U-shaped tube that is connected to a container of gas at one end, open to the atmosphere at the other. The tube contains a liquid that is higher on the left side than on the right. Is the gas pressure p in the container less than, equal to, or greater than 1 atm? Explain.

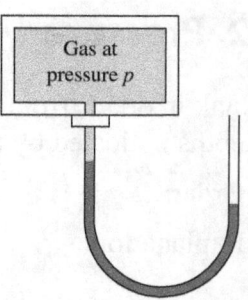

Gas at pressure p

9. It is well known that you can trap liquid in a drinking straw by placing the tip of your finger over the top while the straw is in the liquid, and then lifting it out. The liquid runs out when you release your finger.

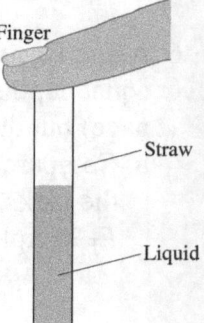

Finger

Straw

Liquid

a. What is the *net* force on the cylinder of trapped liquid?

b. Three forces act on the trapped liquid. Draw and label all three on the figure.
c. Is the gas pressure inside the straw, between the liquid and your finger, greater than, less than, or equal to atmospheric pressure? Explain, basing your explanation on your answers to parts a and b.

d. If your answer to part c was "greater" or "less," how did the pressure change from the atmospheric pressure that was present when you placed your finger over the top of the straw?

13.3 Buoyancy

10. Three blocks of identical size, A, B, and C, are to be gently placed into a large tank of water. Block A has a density of 2 g/cm^3, block B has a density of 0.9 g/cm^3, and block C has a density of 0.5 g/cm^3. On the figure, draw and label each block in its final equilibrium position in the water.

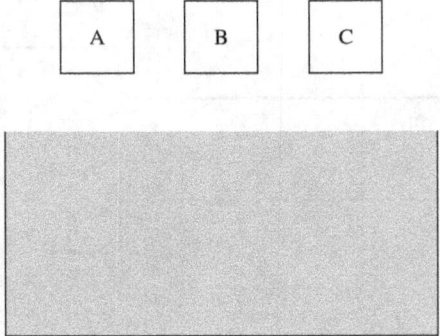

11. Rank in order, from the largest to smallest, the densities of blocks A, B, and C.

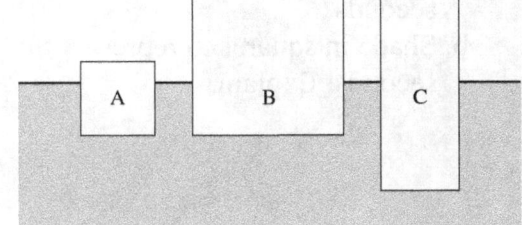

Order:

Explanation:

12. Blocks A, B, and C have the same volume. Rank in order, from largest to smallest, the sizes of the buoyant forces F_A, F_B, and F_C on A, B, and C.

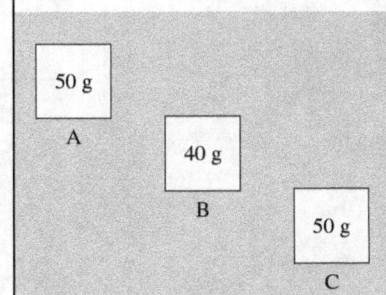

Order:

Explanation:

13.4 Fluids in Motion

13.5 Fluid Dynamics

13. A stream flows from left to right through the constant-depth channel shown below in an overhead view. A 1 m × 1 m grid has been added to facilitate measurement. The fluid's flow speed at A is 2 m/s.

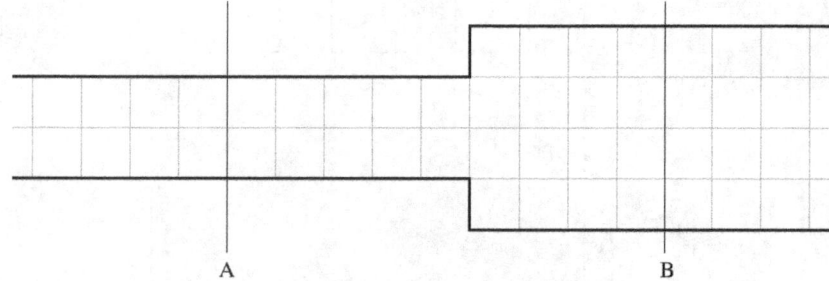

a. Shade in squares to represent the water that has flowed past point A in the last two seconds.

b. Shade in squares to represent the water that has flowed past point B in the last two seconds. Explain.

14. Liquid flows through a tube whose width varies as shown. The liquid level is shown for pipe A, which is open at the top, but not for pipes B and C. Draw an appropriate level of liquid in pipes B and C to indicate the fluid pressures at those points.

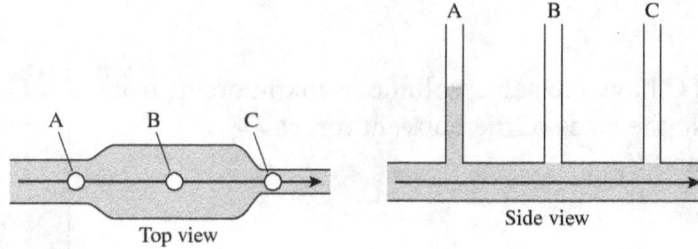

15. Gas flows through a pipe. You can't see into the pipe to know how the inner diameter changes. Rank in order, from largest to smallest, the gas speeds v_1 to v_3 at points 1, 2, and 3.

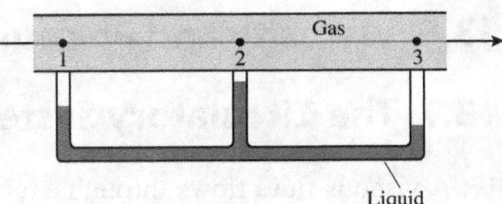

    Order:

    Explanation:

16. Wind blows over a house. A window on the ground floor is open. Is there an air flow through the house? If so, does the air flow in the window and out the chimney, or in the chimney and out the window? Explain.

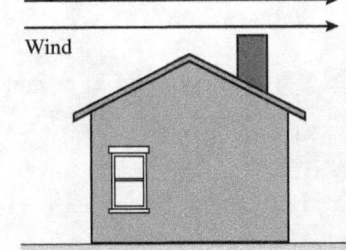

13.6 Viscosity and Poiseuille's Equation

13.7 The Circulatory System

17. A viscous fluid flows through a tube at an average speed of 50 cm/s.
    a. What will be the flow speed if the pressure difference between the ends of the tube is doubled but the tube's diameter is unchanged?

    b. What will be the flow speed if the tube's diameter is doubled but the pressure difference between the ends of the tube is unchanged?

18. A viscous liquid flows through a long tube. Halfway through, at the midpoint, the diameter of the tube suddenly doubles. Suppose the pressure difference between the entrance to the tube and the tube's midpoint is 800 Pa. What is the pressure difference between the midpoint of the tube and the exit?

14 Oscillations

14.1 Equilibrium and Oscillation

14.2 Linear Restoring Forces and SHM

1. On the axes below, sketch three cycles of the position-versus-time graph for:
   a. A particle undergoing simple harmonic motion.

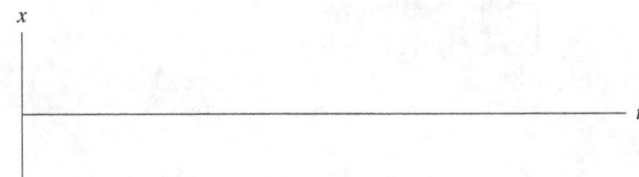

   b. A particle undergoing periodic motion that is not simple harmonic motion.

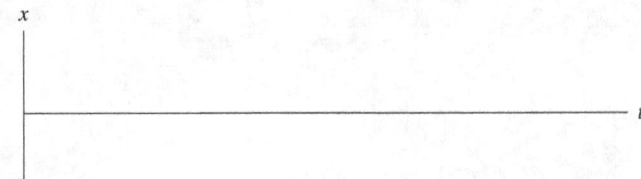

2. Consider the particle whose motion is represented by the x-versus-t graph below.

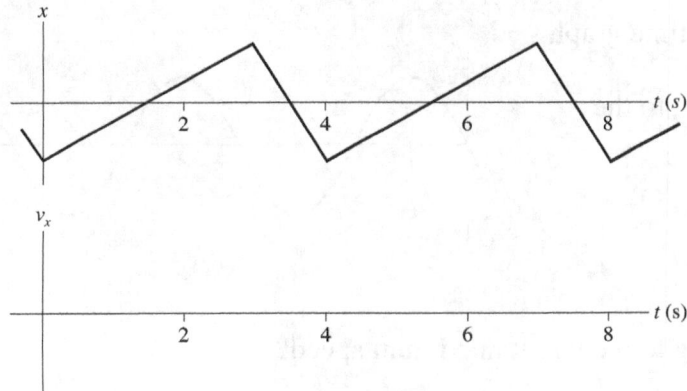

   a. Is this periodic motion? _____   b. Is this motion SHM? _____

   c. What is the period? _____   d. What is the frequency? _____

   e. You learned in Chapter 2 to relate velocity graphs to position graphs. Use that knowledge to draw the particle's velocity-versus-time graph on the axes provided.

3. Figure A shows an unstretched spring. Figure B shows a mass m hanging at rest from the spring. It has stretched the spring by L. Figures C and D show two instants in the oscillation of mass m about the equilibrium position. Draw free-body diagrams for the mass in the Figures B, C, and D.

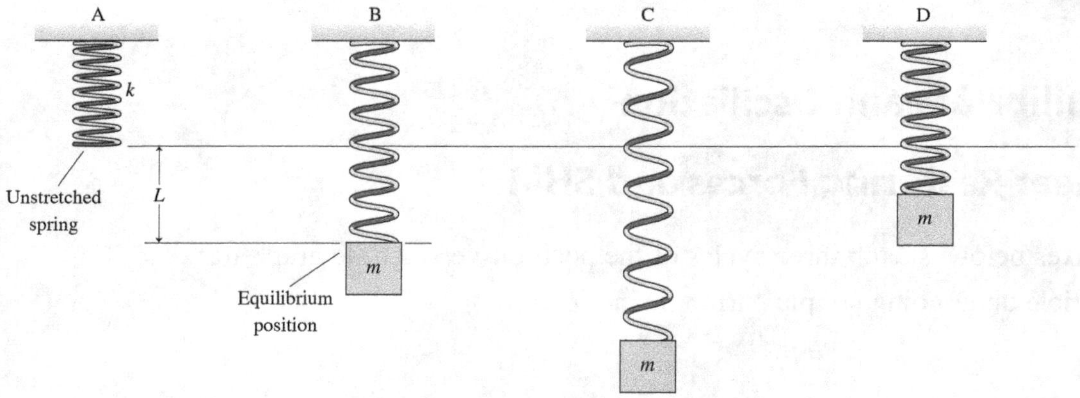

4. The figure shows the position-versus-time graph of a particle in SHM.

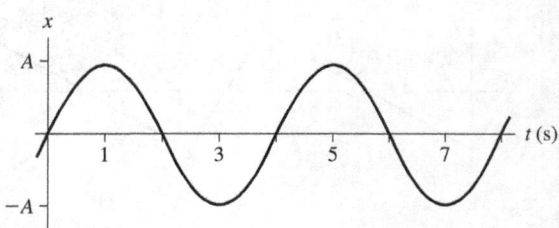

a. At what times is the particle moving to the right at maximum speed?

b. At what times is the particle moving to the left at maximum speed?

c. At what times is the particle instantaneously at rest?

14.3 Describing Simple Harmonic Motion

5. The graph shown is the position-versus-time graph of an oscillating particle.

   a. Draw the corresponding velocity-versus-time graph.

   b. Draw the corresponding acceleration-versus-time graph.

     Hint: Remember that velocity is the slope of the position graph, and acceleration is the slope of the velocity graph.

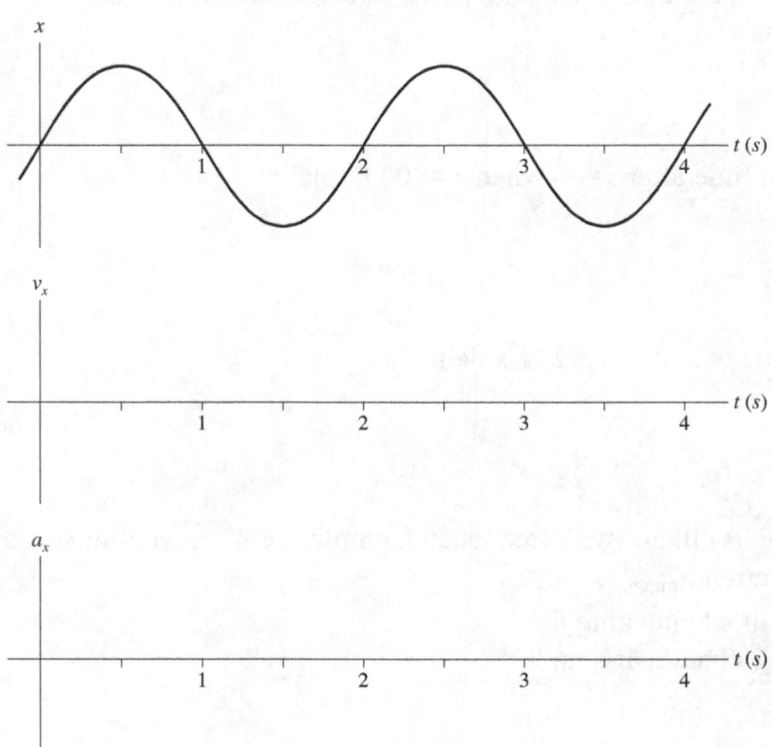

   c. At what times is the position a maximum? _____

     At those times, is the velocity a maximum, a minimum, or zero? _____

     At those times, is the acceleration a maximum, a minimum, or zero? _____

   d. At what times is the position a minimum (most negative)? _____

     At those times, is the velocity a maximum, a minimum, or zero? _____

     At those times, is the acceleration a maximum, a minimum, or zero? _____

   e. At what times is the velocity a maximum? _____

     At those times, where is the particle? _____

   f. Can you find a simple relationship between the *sign* of the position and the *sign* of the acceleration at the same instant of time? If so, what is it?

6. Consider the function $x(t) = A\cos(2\pi t/T)$.

   a. What are the units of the quantity $2\pi t/T$? _____

   b. What is the value of x at $t = 0$? Explain.

   c. What are the next *two* times at which x has the same value as it does at $t = 0$? Your answer will be in terms of the quantity T and, perhaps, constants such as π.

   d. What is the first time after $t = 0$ when $x = 0$? Explain.

   e. What is the value of x at $t = T/2$? Explain.

7. A mass on a spring oscillates with frequency f, amplitude A, maximum speed v_{max}, and maximum acceleration a_{max}.

   a. If A doubles without changing f,

      i. how does v_{max} change, if at all?

      ii. how does a_{max} change, if at all?

   b. If f doubles without changing A,

      i. how does v_{max} change, if at all?

      ii. how does a_{max} change, if at all?

14.4 Energy in Simple Harmonic Motion

8. The figure shows a graph of the potential energy of a block oscillating on a spring. The horizontal line represents the block's total energy E.

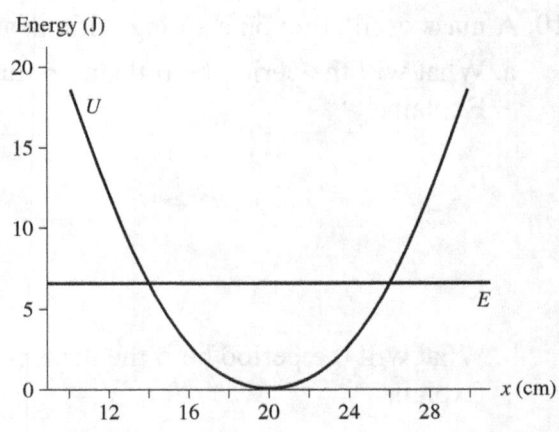

a. What is the spring's equilibrium length?

b. Where are the turning points of the motion? Explain how you identify them.

c. What is the block's maximum kinetic energy?

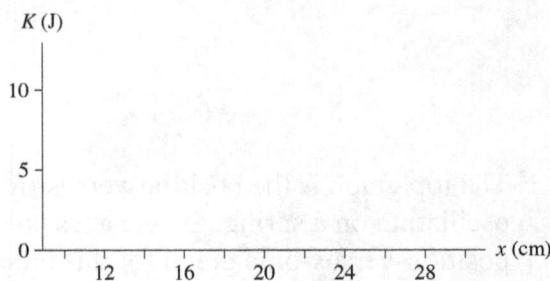

d. Draw a graph of the block's kinetic energy as a function of position.

e. What will be the turning points if the block's total energy is doubled?

9. Equation 14.22 in the textbook states that $\frac{1}{2} m(v_{max})^2 = \frac{1}{2} kA^2$. What does this mean? Write a couple of sentences explaining how to interpret this equation.

10. A mass oscillating on a spring with an amplitude of 5.0 cm has a period of 2.0 s.

    a. What will the period be if the amplitude is doubled to 10.0 cm without changing the mass? Explain.

    b. What will the period be if the mass is doubled without changing the 5.0 cm amplitude? Explain.

11. The top graph is the position-versus-time graph for a mass oscillating on a spring. On the axes below, sketch the position-versus-time graph for this block for the following situations:

    Note: The changes described in each part refer back to the original oscillation, not to the oscillation of the previous part of the question. Assume that all other parameters remain constant. Use the same horizontal and vertical scales as the original oscillation graph.

    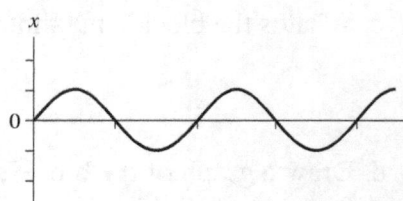

    a. The amplitude and the frequency are both doubled.

    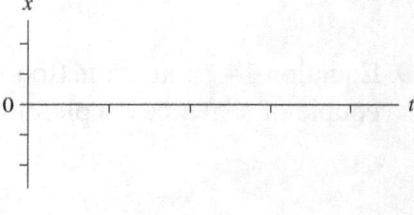

    b. The amplitude is halved and the mass is quadrupled.

    c. The total energy is doubled.

    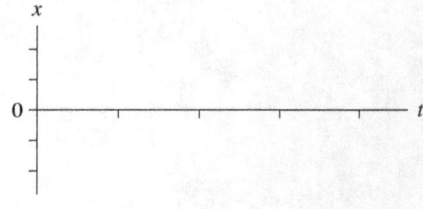

14.5 Pendulum Motion

12. The graph shows the arc-length displacement s versus time for an oscillating pendulum.

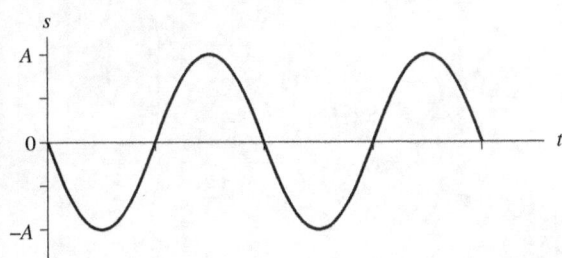 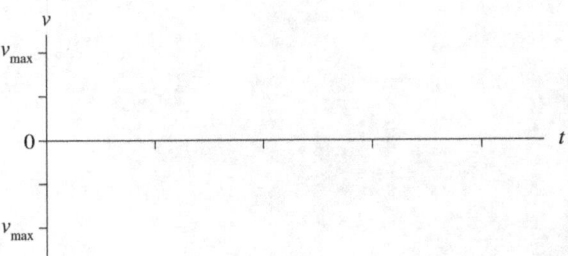

   a. Draw the pendulum's velocity-versus-time graph.

   b. In the space at the right, draw a *picture* of the
      pendulum that shows (and labels!)
      • The extremes of its motion.
      • Its position at $t = 0$ s.
      • Its direction of motion (using an arrow)
        at $t = 0$ s.

13. A pendulum on planet X, where the value of g is unknown, oscillates with a period of
    2 seconds. What is the period of this pendulum if:

   a. Its mass is increased by a factor of 4?
   Note: You do not know the values of m, L, or g, so do not assume any specific values.

   b. Its length is increased by a factor of 4?

   c. Its oscillation amplitude is increased by a factor of 4?

14.6 Damped Oscillations

14. If the time constant τ of an oscillator is decreased, do the oscillations die away more quickly or less quickly? Explain.

15. The figure below shows the decreasing amplitude of a damped oscillator. (The oscillations are occurring rapidly and are not shown; this shows only their amplitude.) On the same axes, draw the amplitude if (a) the time constant is doubled and (b) the time constant is halved. Label your two curves "doubled" and "halved."

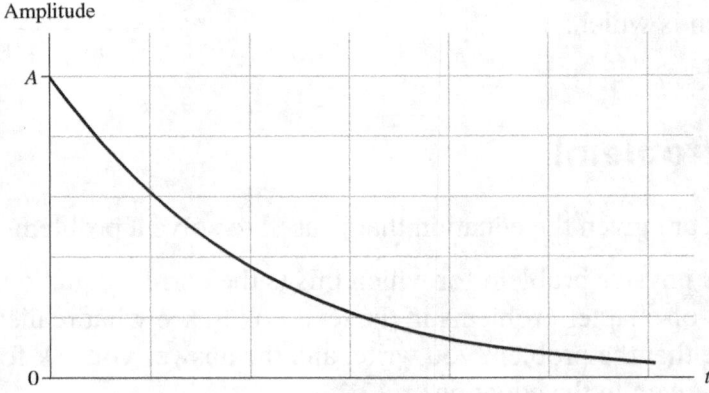

16. For a damped oscillation, is the time constant τ greater than or less than the time in which the oscillation amplitude decays to half of its initial value? Explain.

17. An oscillator has an initial amplitude of 12 cm at $t = 0$ s. The amplitude has decayed to 6 cm at $t = 20$ s. At what time will the amplitude be 3 cm? Explain.

14.7 Driven Oscillations and Resonance

18. A car drives along a bumpy road on which the bumps are equally spaced. At a speed of 20 mph, the frequency of hitting bumps is equal to the natural frequency of the car bouncing on its springs.

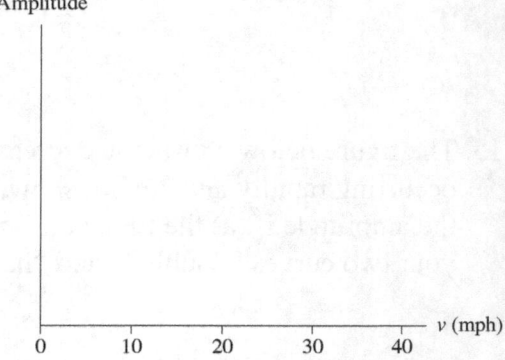

a. Draw a graph of the car's vertical bouncing amplitude as a function of its speed if the car has new shock absorbers (large damping coefficient).

b. Draw a graph of the car's vertical bouncing amplitude as a function of its speed if the car has worn out shock absorbers (small damping coefficient).

Draw both graphs on the same axes, and label them as to which is which.

You Write the Problem!

Exercises 19–20: You are given the equation that is used to solve a problem. For each of these:

a. Write a *realistic* physics problem for which this is the correct equation. Look at worked examples and end-of-chapter problems in the textbook to see what realistic physics problems are like. Be sure that the problem you write, and the answer you ask for, is consistent with the information given in the equation.

b. Finish the solution of the problem.

19. $1.7 \text{ m/s} = \dfrac{2\pi A}{0.75 \text{ s}}$

20. $2.8 \text{ s} = 2\pi \sqrt{\dfrac{1.5 \text{ m}}{g}}$

15 Traveling Waves and Sound

15.1 The Wave Model

15.2 Traveling Waves

1. a. In your own words, define what a *transverse wave* is.

   b. Give an example of a wave that, from your own experience, you know is a transverse wave. What observations or evidence tells you this is a transverse wave?

2. a. In your own words, define what a *longitudinal wave* is.

   b. Give an example of a wave that, from your own experience, you know is a longitudinal wave. What observations or evidence tells you this is a longitudinal wave?

3. Three wave pulses travel along the same stretched string. Rank in order, from largest to smallest, their wave speeds v_1, v_2, and v_3.

   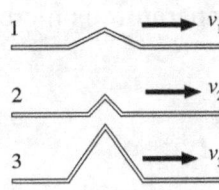

   Order:

   Explanation:

4. A wave pulse travels along a stretched string at a speed of 200 cm/s. What will be the speed if:
   Note: Each part below is independent and refers to changes made to the original string.

   a. The string's tension is doubled?

   b. The string's mass is quadrupled (but its length is unchanged)?

   c. The string's length is quadrupled (but its mass is unchanged)?

   d. The string's mass and length are both quadrupled?

5. Sound travels through a 300 K gas at 400 m/s. What will be the sound speed if the gas temperature is increased to 600 K? Explain.

15.3 Graphical and Mathematical Descriptions of Waves

6. Each figure below shows a snapshot graph at time $t = 0$ s of a wave pulse on a string. The pulse on the left is traveling to the right at 100 cm/s; the pulse on the right is traveling to the left at 100 cm/s. Draw snapshot graphs of the wave pulse at the times shown next to the axes.

a.
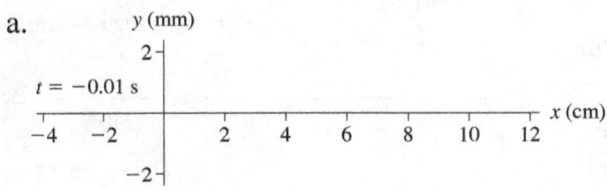
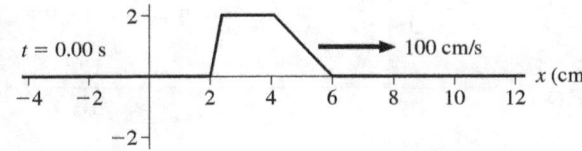
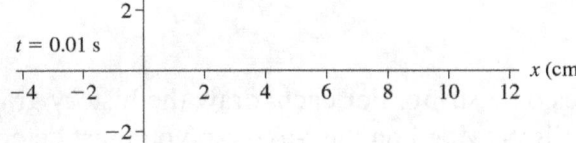
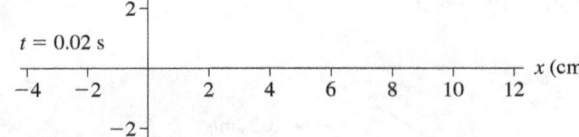
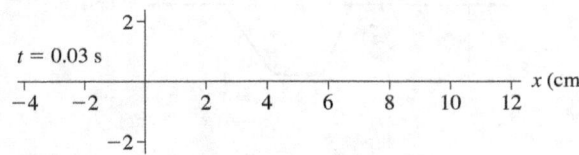
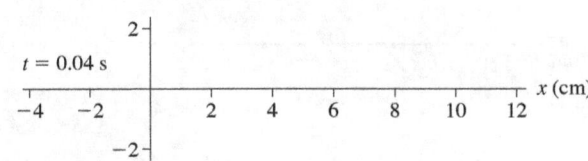

b.
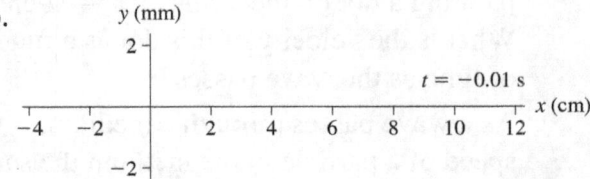
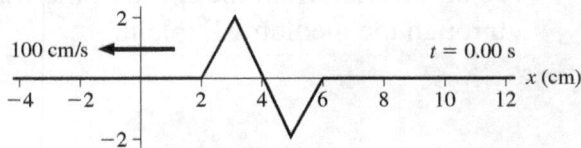
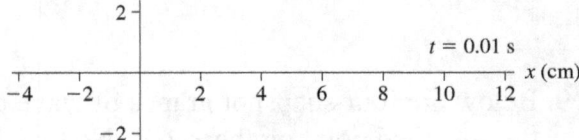
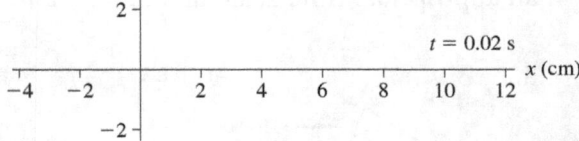
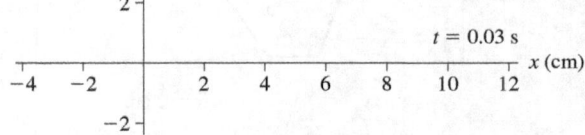
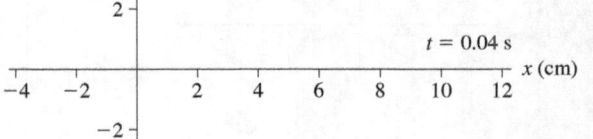

7. This snapshot graph is taken from Exercise 6a. On the axes below, draw the *history* graphs $y(x = 2$ cm, $t)$ and $y(x = 6$ cm, $t)$ showing the displacement at $x = 2$ cm and $x = 6$ cm as functions of time. Refer to your graphs in Exercise 6a to see what is happening at different instants of time.

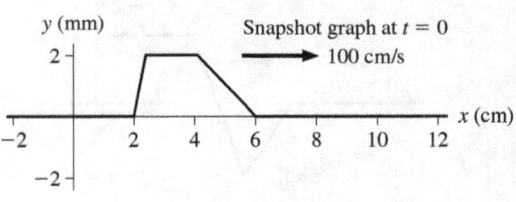

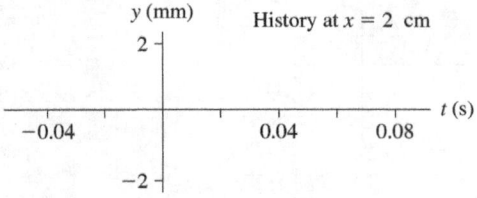

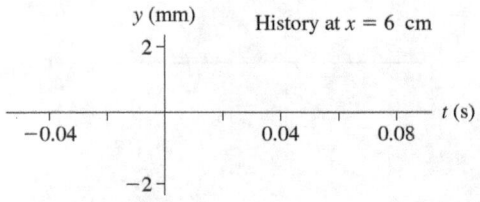

8. This snapshot graph is from Exercise 6b.

   a. Draw the history graph $y(x = 0$ cm, $t)$ for this wave at the point $x = 0$ cm.

   b. Draw the *velocity*-versus-time graph for the piece of the string at $x = 0$ cm. Imagine painting a dot on the string at $x = 0$ cm. What is the velocity of this dot as a function of time as the wave passes by?

   c. As a wave passes through a medium, is the speed of a particle in the medium the same as or different from the speed of the wave through the medium? Explain.

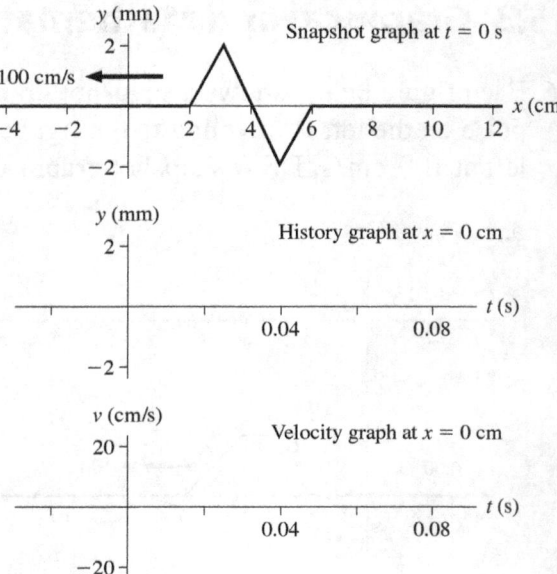

9. Below are four snapshot graphs of wave pulses on a string. For each, draw the history graph at the specified point on the x-axis. No time scale is provided on the t-axis, so you must determine an appropriate time scale and label the t-axis appropriately.

a.
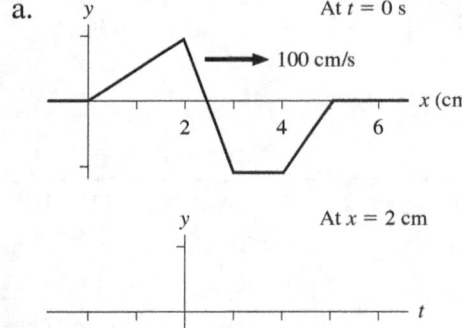

b.
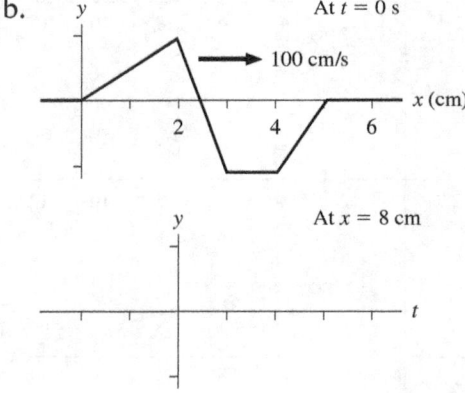

c.
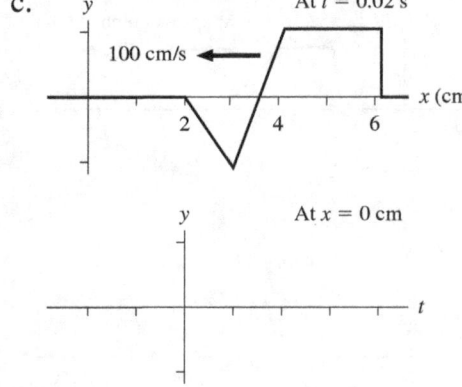

d.
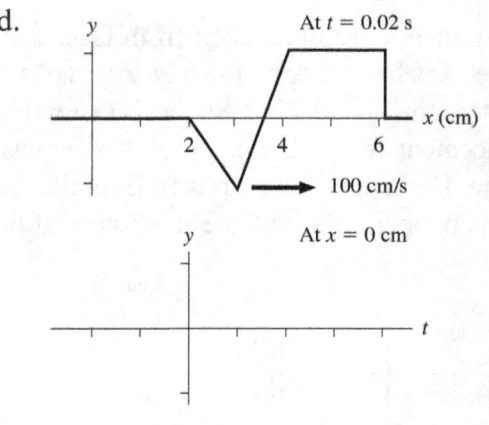

10. A history graph $y(x = 0$ cm, $t)$ is shown for the $x = 0$ cm point on a string. The pulse is moving to the right at 100 cm/s.

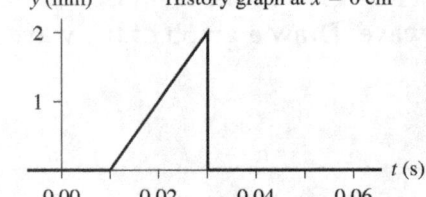

a. Does the $x = 0$ cm point on the string rise quickly and then fall slowly, or rise slowly and then fall quickly? Explain.

b. At what time does the leading edge of the wave pulse arrive at $x = 0$ cm? _____

c. At $t = 0$ s, how far is the leading edge of the wave pulse from $x = 0$ cm? Explain.

d. At $t = 0$ s, is the leading edge to the right or to the left of $x = 0$ cm? _____

e. At what time does the trailing edge of the wave pulse leave $x = 0$ cm? _____

f. At $t = 0$ s, how far is the trailing edge of the pulse from $x = 0$ cm? Explain.

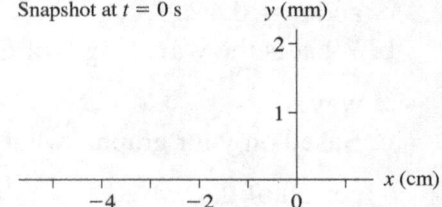

g. By referring to the answers you've just given, draw a snapshot graph $y(x, t = 0$ s) showing the wave pulse on the string at $t = 0$ s.

11. These are a history graph *and* a snapshot graph for a wave pulse on a string. They describe the same wave from two perspectives.

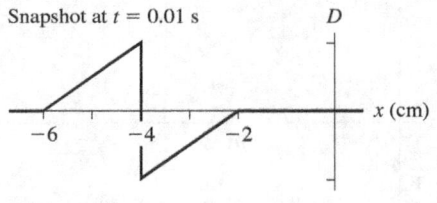

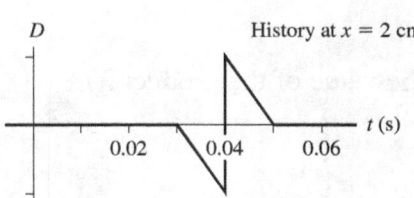

What is the speed of this wave?

12. The figure shows a sinusoidal traveling wave. Draw a graph of the wave if:

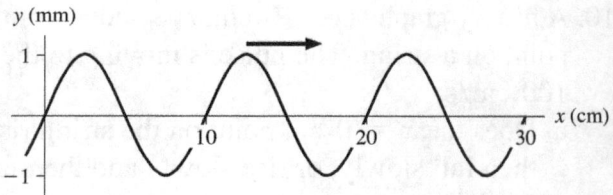

a. Its amplitude is halved and its wavelength is doubled.

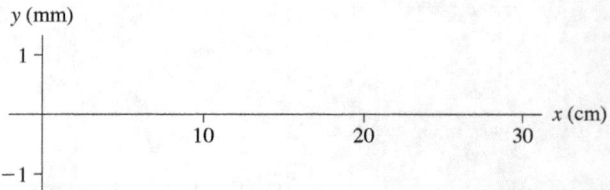

b. Its speed is doubled and its frequency is quadrupled.

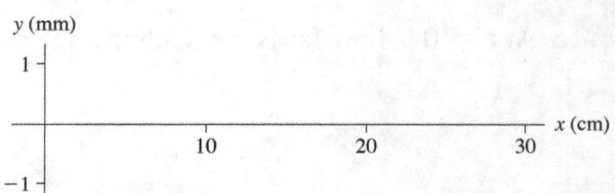

13. The wave shown at time $t = 0$ s is traveling to the right at a speed of 25 cm/s.

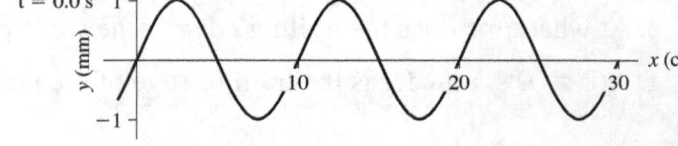

a. Draw snapshot graphs of this wave at times $t = 0.1$ s, $t = 0.2$ s, $t = 0.3$ s, and $t = 0.4$ s.

b. What is the wavelength of the wave? _____

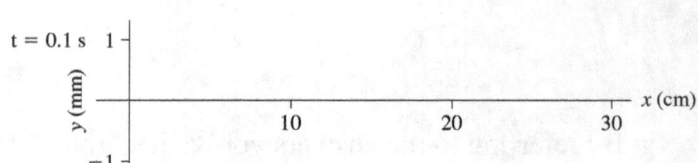

c. Based on your graphs, what is the period of the wave? _____

d. What is the frequency of the wave?
_____

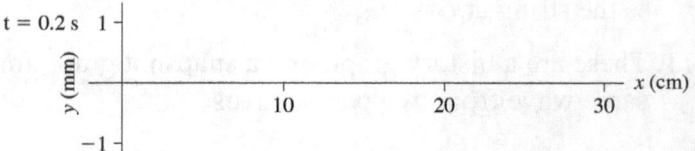

e. What is the value of the product λf?
_____

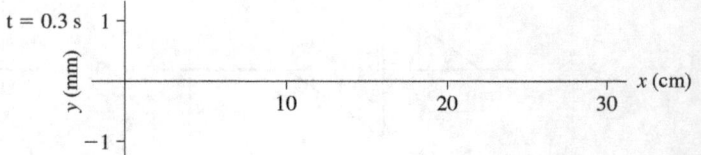

f. How does this value of λf compare to the speed of the wave?
_____

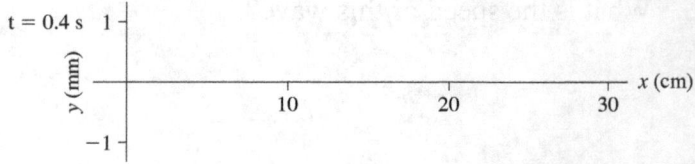

15.4 Sound and Light Waves

14. A horizontal Slinky is at rest on a table. Initially the top of link 5 is at $x = 5.0$ cm. A wave pulse is sent along the Slinky, causing the top of link 5 to move *horizontally* with the *displacement* from equilibrium shown in the graph.

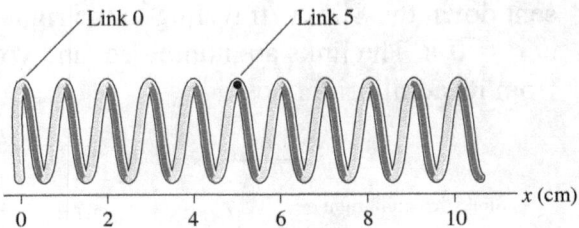

a. Is this a transverse or a longitudinal wave? Explain.

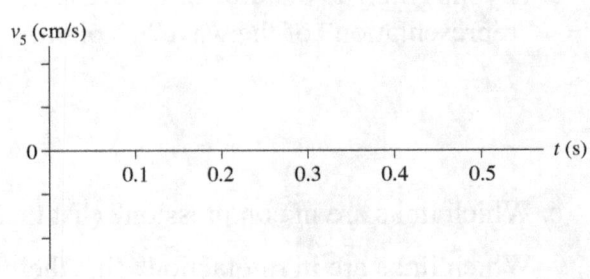

b. What is the *position* of link 5

at $t = 0.1$ s? _____

What is the *position* of link 5

at $t = 0.2$ s? _____

What is the *position* of link 5

at $t = 0.3$ s? _____

Note: *Position*, not displacement.

c. Draw a velocity-versus-time graph of link 5. Add an appropriate scale to the vertical axis. (Recall how velocity graphs are related to the slopes of position graphs.)

d. Can you determine, from the information given, the speed of the wave? If so, give the speed and explain how you found it. If not, why not?

15. We can use a series of dots to represent the positions of the links in a Slinky. The top set of dots shows a Slinky in equilibrium with a 1 cm spacing between the links. A wave pulse is sent down the Slinky, traveling to the right at 10 cm/s. The second set of dots show the Slinky at $t = 0$ s. The links are numbered, and you can measure the displacement Δx of each link from its equilibrium position.

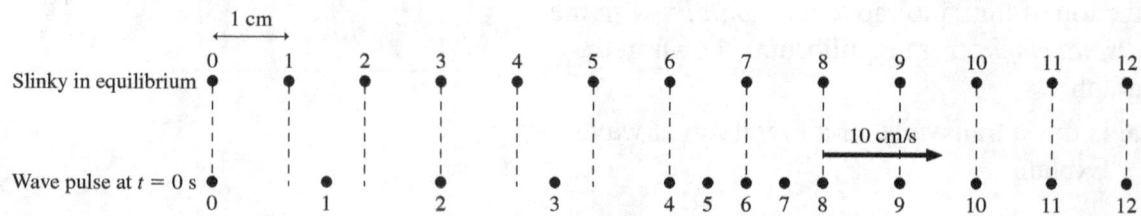

a. Draw a snapshot graph showing the displacement Δx of each link at $t = 0$ s. There are 13 links, so your graph should have 13 dots. Connect your dots with lines to make a continuous graph.

b. Is your graph a "picture" of the wave or a "representation" of the wave? Explain.

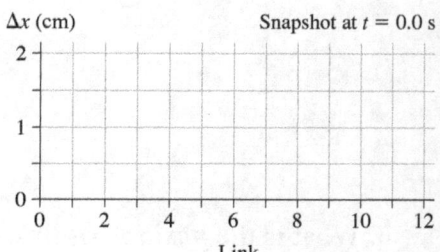

c. Which links are in compression? (list their numbers) _____

Which links are in rarefaction? (list their numbers) _____

d. Draw graphs of displacement versus the link number at $t = 0.2$ s and $t = 0.4$ s.

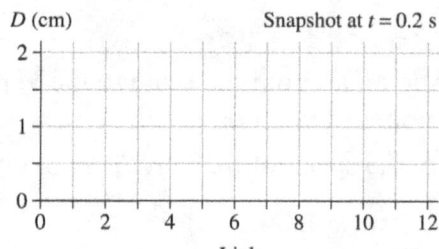 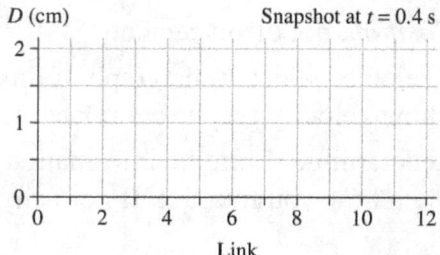

16. Rank in order, from largest to smallest, the wavelengths λ_1 to λ_3 for sound waves having frequencies $f_1 = 100$ Hz, $f_2 = 1000$ Hz, and $f_3 = 10{,}000$ Hz.

Order:

Explanation:

15.5 Energy and Intensity

15.6 Loudness of Sound

17. The figure shows the path of a light wave past a series of equally spaced *transparent* grids. The portion of the first two grids that would be illuminated by the light is represented by the shaded area in the first two grids.

    a. Complete the figure by shading in the spaces that would be illuminated in the remaining two grids.

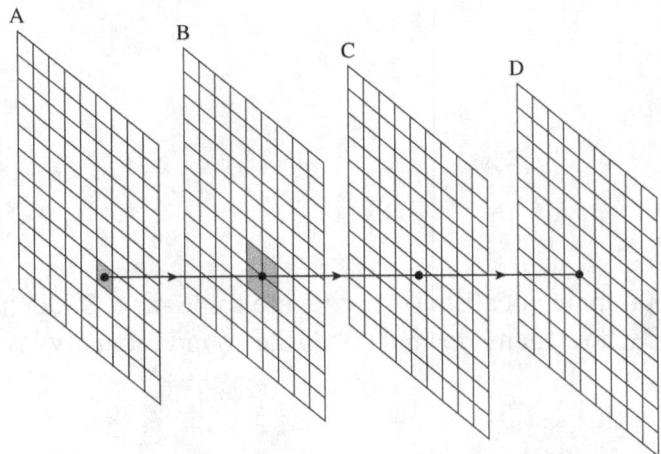

    b. Is the energy of the light wave passing through the fourth transparent grid (D) greater than, less than, or equal to the energy passing through the first transparent grid (A)? Explain.

    c. What is the ratio I_D/I_A of the intensity of the light at the fourth grid (D) to the intensity at the first grid (A)? Explain.

18. A laser beam has intensity I_0.

    a. What is the intensity, in terms of I_0, if a lens focuses the laser beam to $\frac{1}{10}$ its initial diameter?

    b. What is the intensity, in terms of I_0, if a lens defocuses the laser beam to 10 times its initial diameter?

19. Sound wave A delivers 2 J of energy in 2 s. Sound wave B delivers 10 J of energy in 5 s. Sound wave C delivers 2 mJ of energy in 1 ms. Rank in order, from largest to smallest, the sound powers P_A, P_B, and P_C of these three sound waves.

Order:

Explanation:

20. A giant chorus of 1000 male vocalists is singing the same note. Suddenly, 999 vocalists stop, leaving one soloist. By how many decibels does the sound intensity level decrease? Explain.

15.7 The Doppler Effect and Shock Waves

21. Five expanding wave fronts from a moving sound source are shown. The dots represent the centers of the respective circular wave fronts, which is the location of the source when that wave front was emitted. The frequency of the sound emitted by the source is constant.

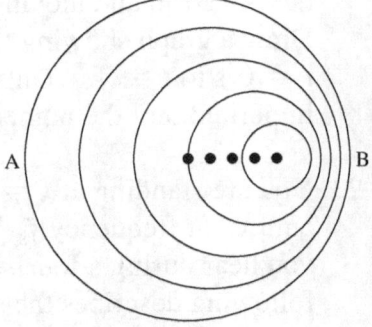

a. Indicate on the figure the direction of motion of the source. Which sound wave front was produced first? How do you know? Explain.

b. Do the observers at locations A and B hear the same frequency of sound? If not, which one hears a higher frequency? Explain.

c. Assume that the sound wave you identified in part a as the first wave front produced marks the beginning of the sound. Do the observers at A and B first hear the sound at the same time? If not, which one hears the sound first? Explain.

d. The speed of sound in the medium is v. Is the speed v_s of the source greater than, less than, or equal to v? Explain.

22. You are standing at $x = 0$ m, listening to a sound that is emitted at a frequency f_s. At $t = 0$ s, the sound source is at $x = 20$ m and moving towards you at a steady 10 m/s. Draw a graph showing the frequency you hear from $t = 0$ s to $t = 4$ s. Only the shape of the graph is important, not the numerical values of f.

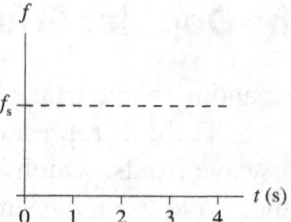

23. You are standing at $x = 0$ m, listening to a sound that is emitted at frequency f_s. The graph shows the frequency you hear during a four-second interval. Which of the following describes the sound source?

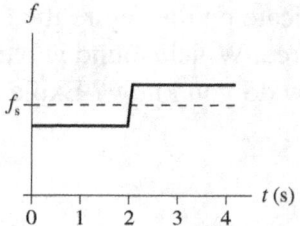

   i.   It moves from left to right and passes you at $t = 2$ s.

   ii. It moves from right to left and passes you at $t = 2$ s.

   iii. It moves towards you but doesn't reach you. It then reverses direction at $t = 2$ s.

   iv. It moves away from you until $t = 2$ s. It then reverses direction and moves towards you but doesn't reach you.

   Explain your choice.

24. You are standing at $x = 0$ m, listening to seven identical sound sources. At $t = 0$ s, all seven are at $x = 343$ m and moving as shown below. The sound from all seven will reach your ear at $t = 1$ s.

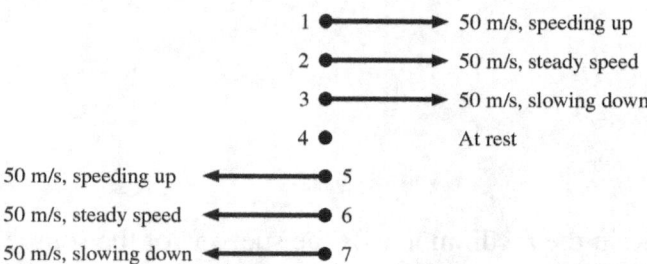

   Rank in order, from highest to lowest, the seven frequencies f_1 to f_7 that you hear at $t = 1$ s.

   Order:

   Explanation:

16 Superposition and Standing Waves

16.1 The Principle of Superposition

1. Two pulses on a string are approaching each other at 10 m/s. Draw snapshot graphs of the string at the three times indicated.

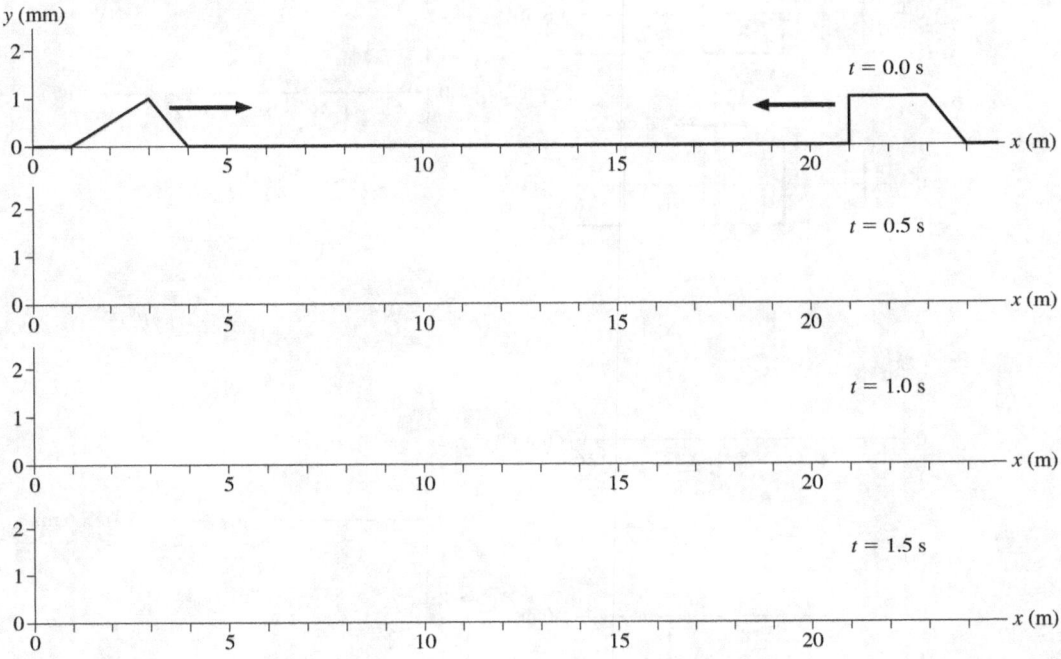

2. Two pulses on a string are approaching each other at 10 m/s. Draw a snapshot graph of the string at $t = 1$ s.

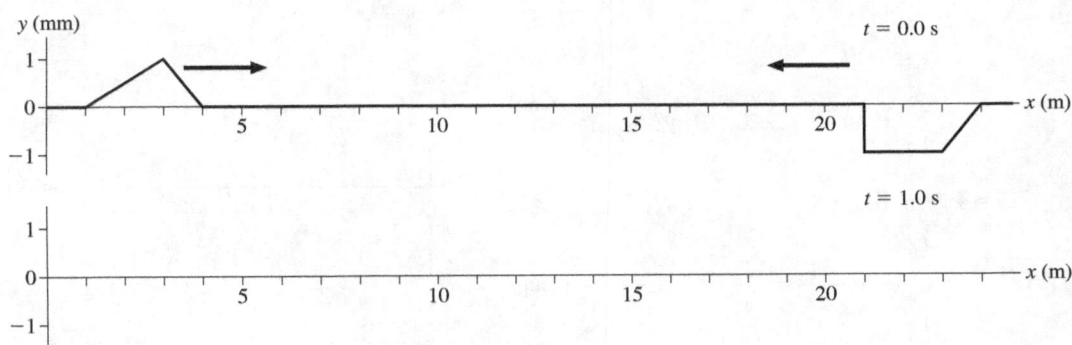

16.2 Standing Waves

3. Two waves are traveling in opposite directions along a string. Each has a speed of 1 cm/s, and an amplitude of 1 cm. The first set of graphs below shows each wave at $t = 0$ s.

   a. On the axes at the right, draw the superposition of these two waves at $t = 0$ s.

   b. On the axes at the left, draw each of the two displacements every 2 s until $t = 8$ s. The waves extend beyond the graph edges, so new pieces of the wave will move in.

   c. On the axes at the right, draw the superposition of the two waves at the same instant.

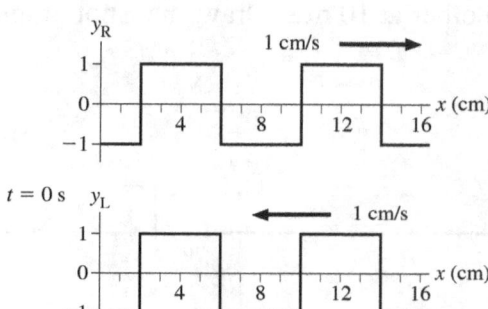
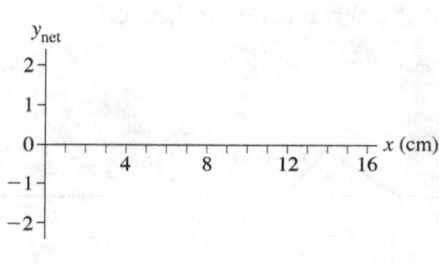

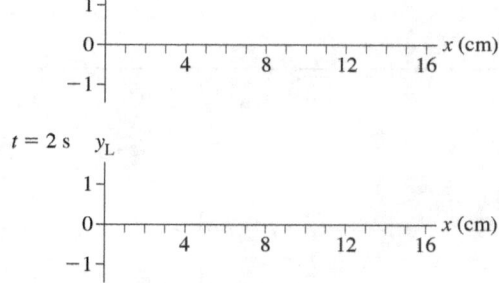
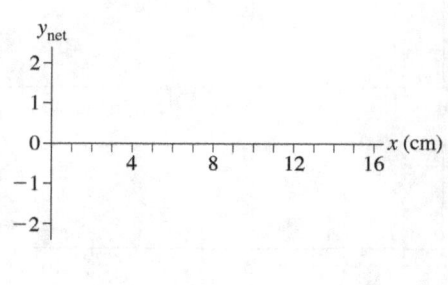

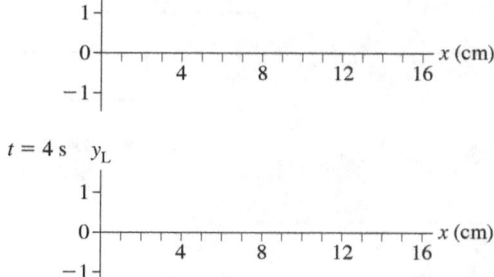
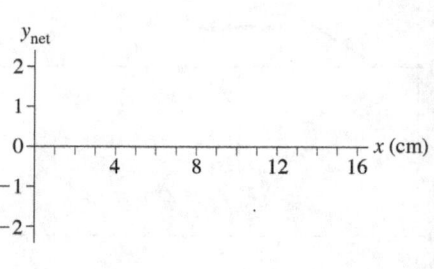

(Continues next page)

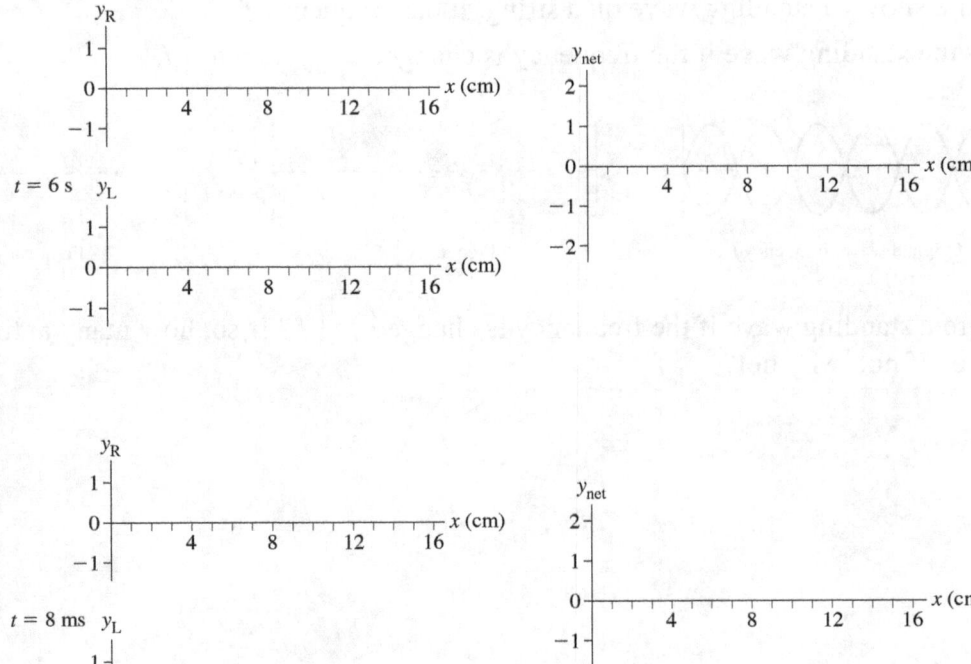

16.3 Standing Waves on a String

4. This standing wave has a period of 8 ms. Draw snapshot graphs of the string every 1 ms from $t = 1$ ms to $t = 8$ ms. Think carefully about the proper amplitude at each instant.

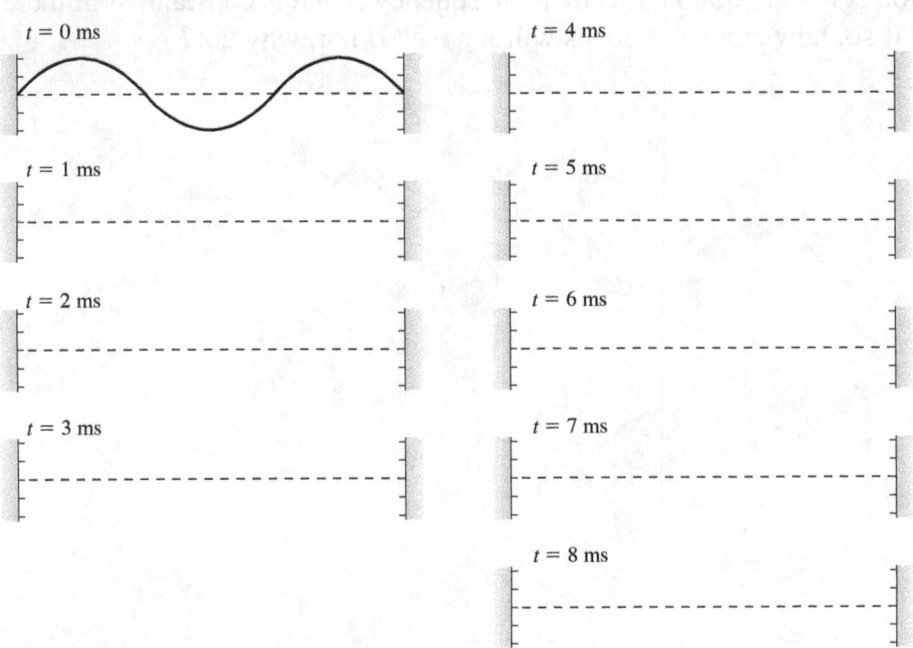

5. The figure shows a standing wave on a string. It has frequency f.

   a. Draw the standing wave if the frequency is changed to $\frac{2}{3}f$ and to $\frac{3}{2}f$.

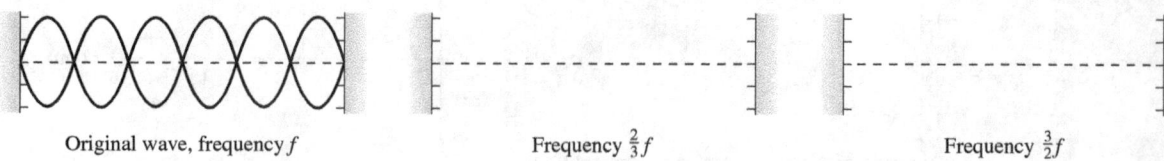

Original wave, frequency f Frequency $\frac{2}{3}f$ Frequency $\frac{3}{2}f$

   b. Is there a standing wave if the frequency is changed to $\frac{1}{4}f$? If so, how many antinodes does it have? If not, why not?

6. The figure shows a standing wave on a string.

   a. Draw the standing wave if the tension is quadrupled while the frequency is held constant.

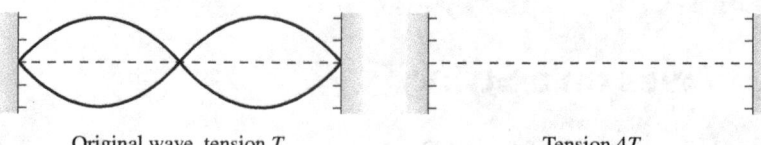

Original wave, tension T Tension $4T$

   b. Suppose the tension is merely doubled while the frequency remains constant. Will there be a standing wave? If so, how many antinodes will it have? If not, why not?

16.4 Standing Sound Waves

7. The picture shows a *displacement* graph (not a pressure graph) of a standing sound wave in a 32-mm-long horizontal tube of air that is open at both ends. That is, the graph shows how far a molecule is displaced from its equilibrium position.

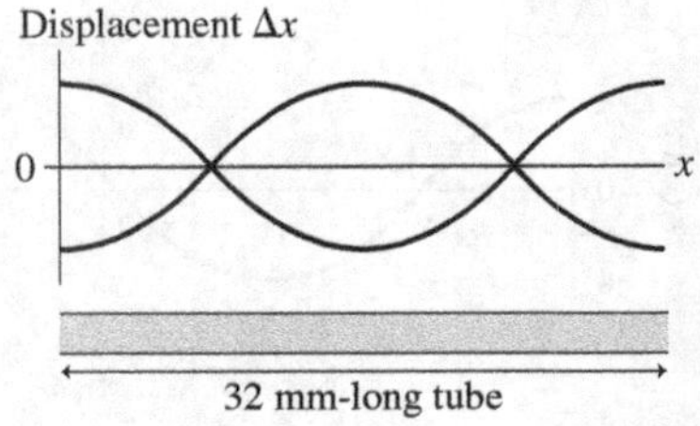

   Displacement Δx

   0 ——— x

   32 mm-long tube

   a. Which mode (value of m) standing wave is this?

   b. Are the air molecules vibrating vertically or horizontally? Explain.

   c. At what distances from the left end of the tube do the molecules oscillate with maximum amplitude?

8. The purpose of this exercise is to visualize the motion of the air molecules for the standing wave of Exercise 7. On the next page are nine graphs, every one-eighth of a period from $t = 0$ to $t = T$. Each graph represents the displacements at that instant of time of the molecules in a 32-mm-long tube. Positive values are displacements to the right, negative values are displacements to the left.

   a. Consider nine air molecules that, in equilibrium, are 4 mm apart and lie along the axis of the tube. The top picture on the right shows these molecules in their equilibrium positions. The dotted lines down the page—spaced 4 mm apart—are reference lines showing the equilibrium positions. Read each graph carefully, then draw nine dots to show the positions of the nine air molecules at each instant of time. The first one, for $t = 0$, has already been done to illustrate the procedure.

   Note: It's a good approximation to assume that the left dot moves in the pattern 4, 3, 0, −3, −4, −3, 0, 3, 4 mm; the second dot in the pattern 3, 2, 0, −2, −3, −2, 0, 2, 3 mm; and so on.

   b. At what times does the air reach maximum compression, and where does it occur?

   Max compression at time _____   Max compression at position _____

   c. What is the relationship between the positions of maximum compression and the nodes of the standing wave?

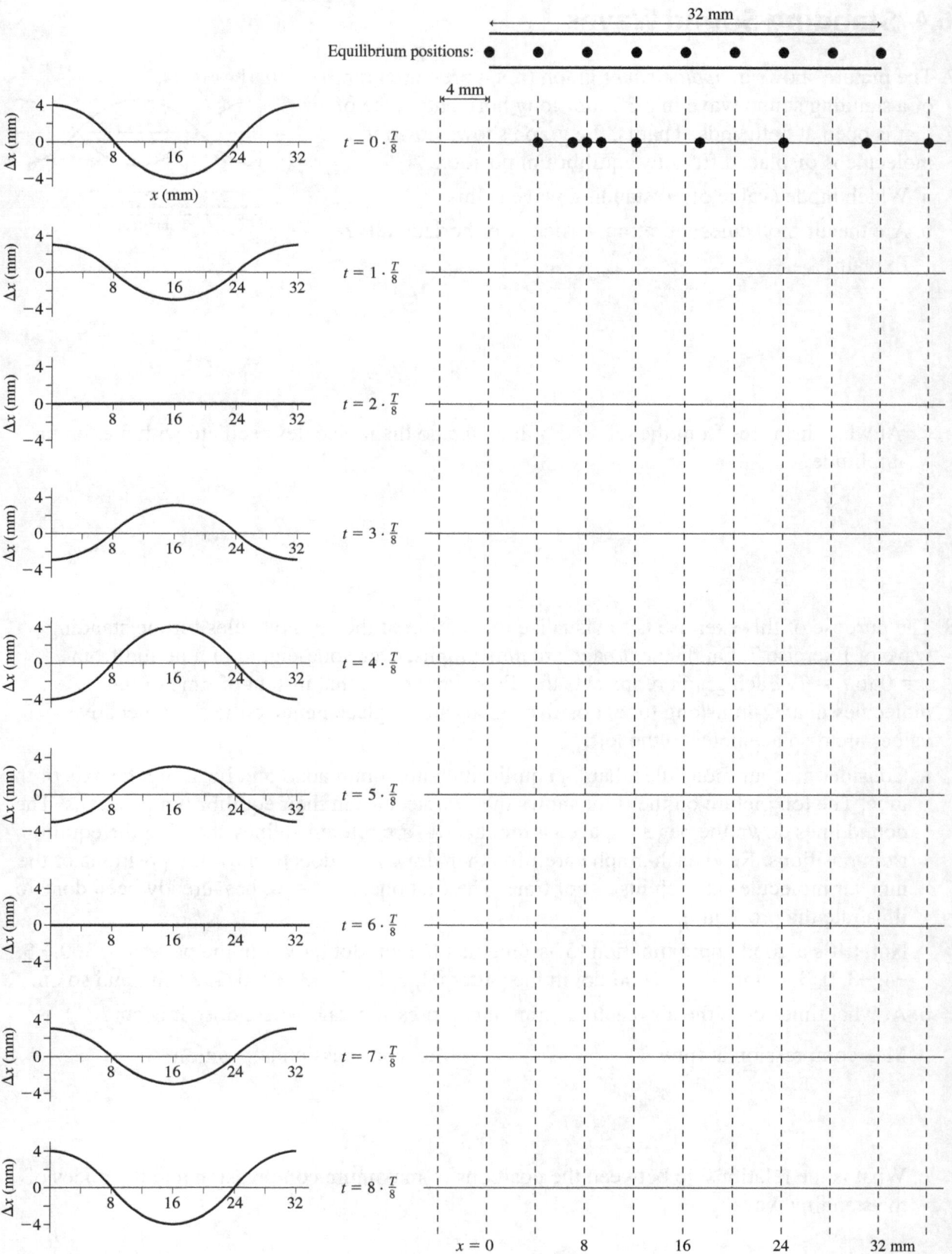

16.5 Speech and Hearing

16.6 The Interference of Waves from Two Sources

9. The figure shows a snapshot graph at $t = 0$ s of loudspeakers emitting triangular-shaped sound waves. Speaker 2 can be moved forward or backward along the axis. Both speakers vibrate in phase at the same frequency. The second speaker is drawn below the first, so that the figure is clear, but you want to think of the two waves as overlapped as they travel along the x-axis.

   a. On the left set of axes, draw the $t = 0$ s snapshot graph of the second wave if speaker 2 is placed at each of the positions shown. The first graph, with $x_{\text{speaker}} = 2$ m, is already drawn.

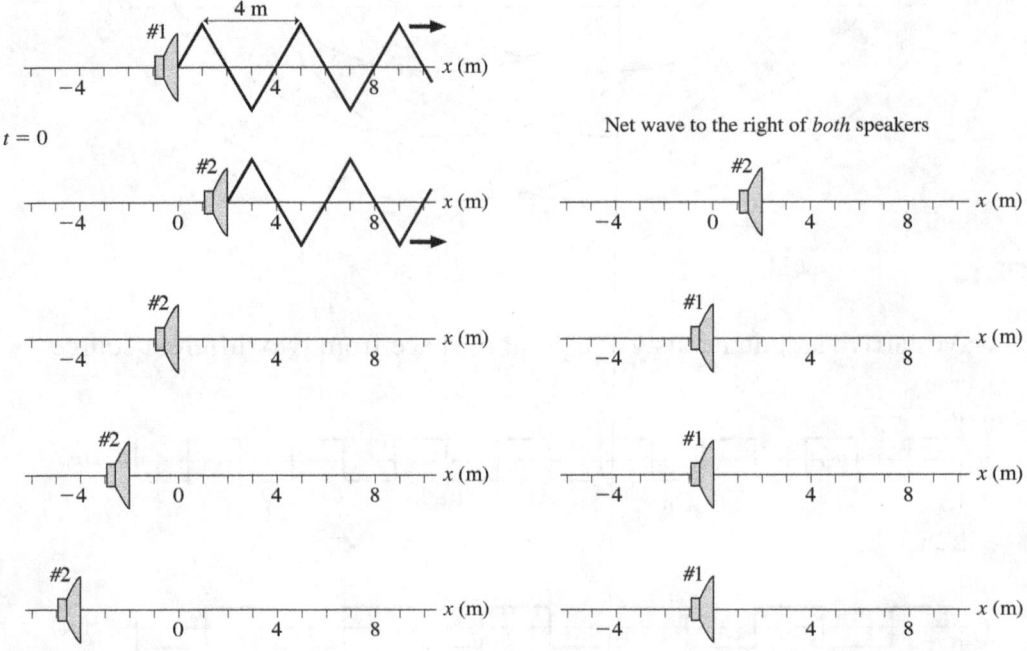

   b. On the right set of axes, draw the superposition $\Delta p = \Delta p_1 + \Delta p_2$ of the waves from the two speakers. Δp exists only to the right of *both* speakers. It is the net wave traveling to the right.

   c. What distance or distances Δd between the speakers give constructive interference?

   d. What are the values of $\Delta d / \lambda$, the ratio of path-length distance to wavelength, at the points of constructive interference?

   e. What distance or distances Δd between the speakers give destructive interference?

   f. What are the values of $\Delta d / \lambda$ at the points of destructive interference?

10. The figure shows the wave-front pattern emitted by two loudspeakers.

    a. Draw a dot • at points where there is constructive interference. These are points where two crests overlap *or* two troughs overlap.

    b. Draw an open circle ○ at points where there is destructive interference. These are points where a crest overlaps a trough.

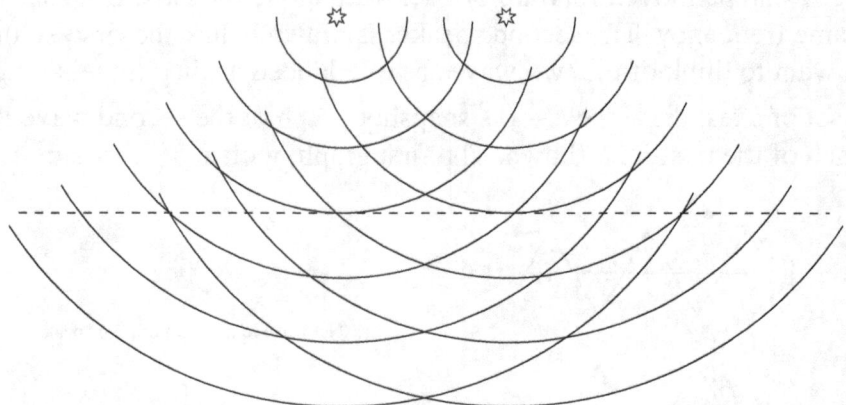

16.7 Beats

11. The two waves arrive simultaneously at a point in space from two different sources.

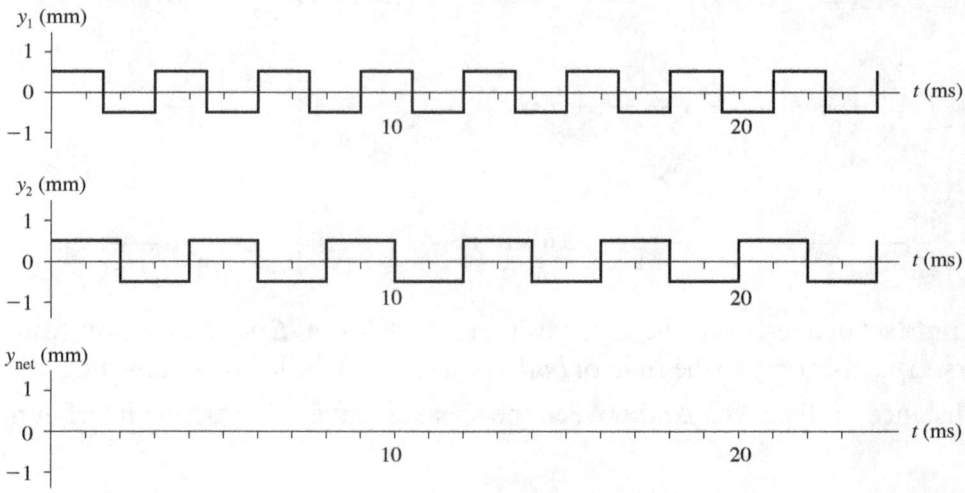

   a. Period of wave 1? _____ Frequency of wave 1? _____

   b. Period of wave 2? _____ Frequency of wave 2? _____

   c. Draw the graph of the net wave at this point on the third set of axes. Be accurate, use a ruler!

   d. Period of the net wave? _____ Frequency of the net wave? _____

   e. Is the frequency of the superposition what you would expect as a beat frequency? Explain.

17 Wave Optics

17.1 What is Light?

1. A light wave travels from vacuum, through a transparent material, and back to vacuum. What is the index of refraction of this material? Explain.

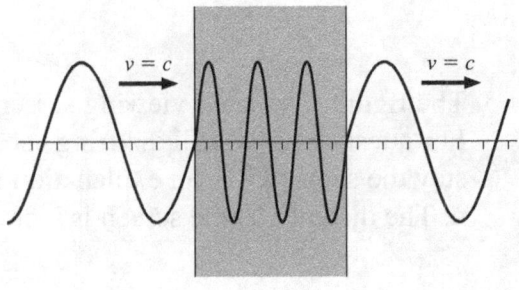

2. A light wave travels from vacuum, through a transparent material whose index of refraction is $n = 2.0$, and back to vacuum. Finish drawing the snapshot graph of the light wave at this instant.

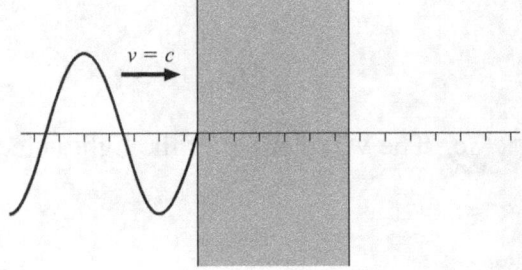

17.2 The Interference of Light

3. The figure shows the light intensity recorded by a detector in an interference experiment. Notice that the light intensity comes "full on" at the edges of each maximum, so this is *not* the intensity that would be recorded in Young's double-slit experiment.

a. Draw a graph of light intensity versus position on the film. Your graph should have the same horizontal scale as the "photograph" above it.

b. Is it possible to tell, from the information given, what the wavelength of the light is? If so, what is it? If not, why not?

4. The graph shows the light intensity on the viewing screen during a double-slit interference experiment. Draw the "photograph" that would be recorded if a light detector were placed at the position of the screen. Your "photograph" should have the same horizontal scale as the graph above it. Be as accurate as you can. Let the white of the paper be the brightest intensity and a very heavy pencil shading be the darkest.

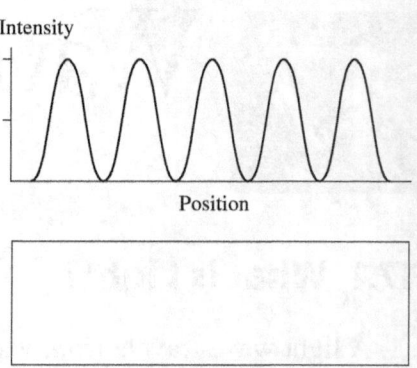

Intensity

Position

Photograph

5. The figure shows the viewing screen in a double-slit experiment. For questions a–c, will the fringe spacing increase, decrease, or stay the same? Give an explanation for each.

a. The distance to the screen is increased.

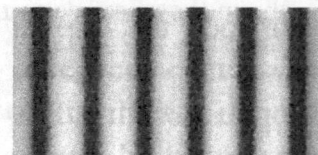

b. The spacing between the slits is increased.

c. The wavelength of the light is increased.

6. In a double-slit experiment, we usually see the light intensity on a viewing screen. However, we can use smoke to make the light visible as it propagates between the slits and the screen. Consider a double-slit experiment in a smoke-filled room. What kind of light and dark pattern would you see if you looked down on the experiment from above? Draw the pattern on the figure below. Shade the areas that are dark and leave the white of the paper for the bright areas. **Hint:** What is the condition for constructive interference? For destructive?

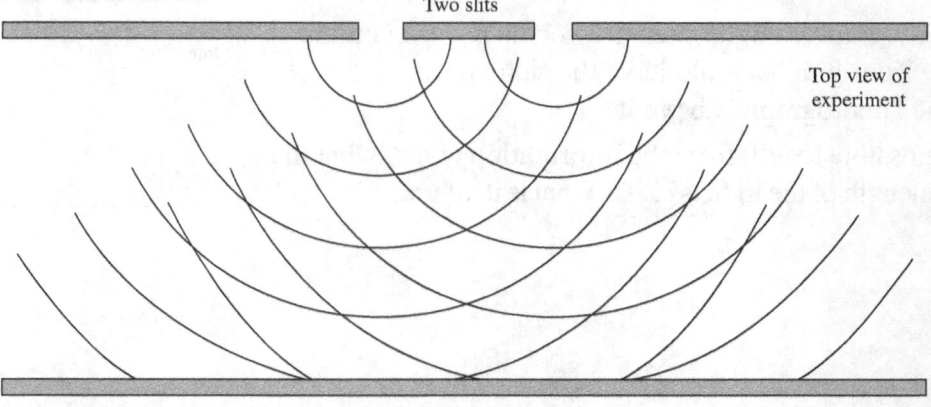

Two slits

Top view of experiment

Viewing screen

17.3 The Diffraction Grating

7. The figure shows four slits in a diffraction grating. A set of circular wave crests is shown spreading out from each slit. Four wave paths, numbered 1 to 4, are shown leaving the slits at angle θ_1. The dotted lines are drawn perpendicular to the paths of the waves.

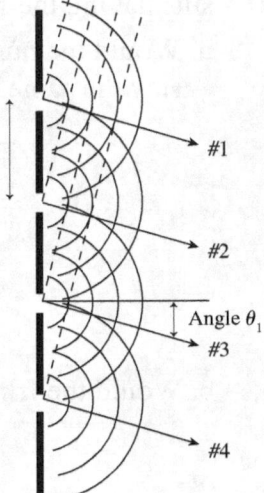

a. Use a colored pencil or heavy shading to show *on the figure* the extra distance traveled by wave 1 that is not traveled by wave 2.

b. How many extra wavelengths does wave 1 travel compared to wave 2? Explain how you can tell from the figure.

c. How many extra wavelengths does wave 2 travel compared to wave 3?

d. As these four waves combine at some large distance from the grating, will they interfere constructively, destructively, or in between? Explain.

8. Suppose the wavelength of the light in Exercise 7 is doubled. (Imagine erasing every other wave front in the picture.) Would the interference at angle θ_1 then be constructive, destructive, or in between? Explain. Your explanation should be based on the figure, not on some equation.

9. Suppose the slit spacing d in Exercise 7 is doubled while the wavelength is unchanged. Would the interference at angle θ_1 then be constructive, destructive, or in between? Again, base your explanation on the figure.

10. This is the interference pattern on a viewing screen behind 2 slits. How would the pattern change if the 2 slits were replaced by 20 slits having the *same spacing d* between adjacent slits?

    a. Would the number of fringes on the screen increase, decrease, or stay the same?

    b. Would the fringe spacing increase, decrease, or stay the same?

    c. Would the width of each fringe increase, decrease, or stay the same?

    d. Would the brightness of each fringe increase, decrease, or stay the same?

17.4 Thin-Film Interference

11. The figure shows a wave transmitted from air through a thin oil film on water. The film has a thickness $t = \lambda_{oil}/2$, where λ_{oil} is the wavelength of the light while in the oil.

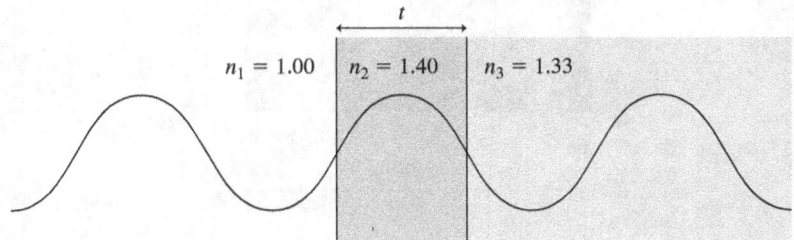

$n_1 = 1.00$ $n_2 = 1.40$ $n_3 = 1.33$

   a. Referring to the indices of refraction shown on the figure, indicate at each boundary with a Y (yes) or N (no) whether the reflected wave undergoes a phase change at the boundary.

   b. Notice the indices of refraction shown on the figure. At each of the two boundaries, write at the bottom of the figure a Y (yes) or N (no) to indicate whether the reflected wave undergoes a phase change at that boundary.

   c. Do the two reflected waves interfere constructively, destructively, or in between? Explain.

12. The figure shows a wave transmitted from air through a thin oil film on glass. The film has a thickness $t = \lambda_{oil}/2$, where λ_{oil} is the wavelength of the light while in the oil.

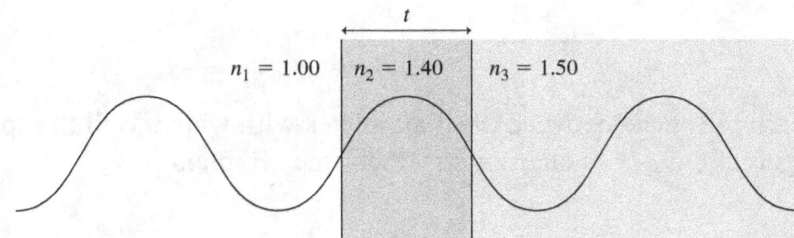

$n_1 = 1.00$ $n_2 = 1.40$ $n_3 = 1.50$

   a. Referring to the indices of refraction shown on the figure, indicate at each boundary with a Y (yes) or N (no) whether the reflected wave undergoes a phase change at the boundary.

   b. Notice the indices of refraction shown on the figure. At each of the two boundaries, write at the bottom of the figure a Y (yes) or N (no) to indicate whether the reflected wave undergoes a phase change at that boundary.

   c. Do the two reflected waves interfere constructively, destructively, or in between? Explain.

13. The figure shows the fringes seen due to a wedge of air between two flat glass plates that touch at one end and are illuminated by light of wavelength $\lambda = 500$ nm.

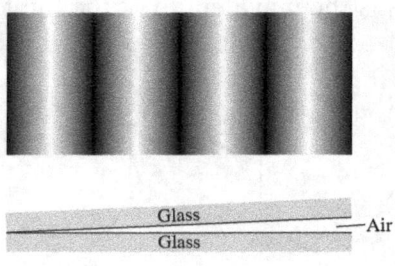

a. By how much does the wedge of air increase in thickness as you move from one dark fringe to the next dark fringe? Explain.

b. By how much does the wedge of air increase in thickness from one end of the above figure to the other?

c. Suppose you fill the space between the glass plates with water. Will the spacing between the dark fringes get larger, get smaller, or stay the same? Explain.

17.5 Single-Slit Diffraction

14. Plane waves of light are incident on two narrow, closely-spaced slits. The graph shows the light intensity seen on a screen behind the slits.

    a. Draw a graph on the bottom axes to show the light intensity on the screen if the right slit is blocked, allowing light to go only through the left slit.

    b. Explain why the graph will look this way.

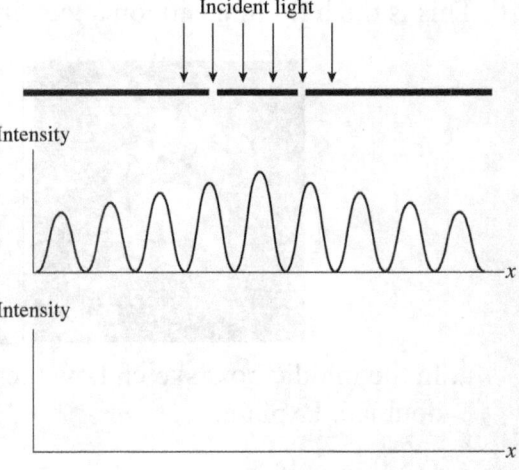

15. The graph shows the light intensity on a screen behind a 0.2-mm-wide slit illuminated by light with a 500 nm wavelength.

    a. Draw a *picture* in the box of how a photograph taken at this location would look. Use the same horizontal scale, so that your picture aligns with the graph above. Let the white of the paper represent the brightest intensity and the darkest, you can draw with a pencil or pen, be the least intensity.

    b. Using the same horizontal scale as in part a, draw graphs showing the light intensity if

    i. $\lambda = 250$ nm, $a = 0.2$ mm.

    ii. $\lambda = 1000$ nm, $a = 0.2$ mm.

    iii. $\lambda = 500$ nm, $a = 0.1$ mm.

    iv. $\lambda = 500$ nm, $a = 0.4$ mm.

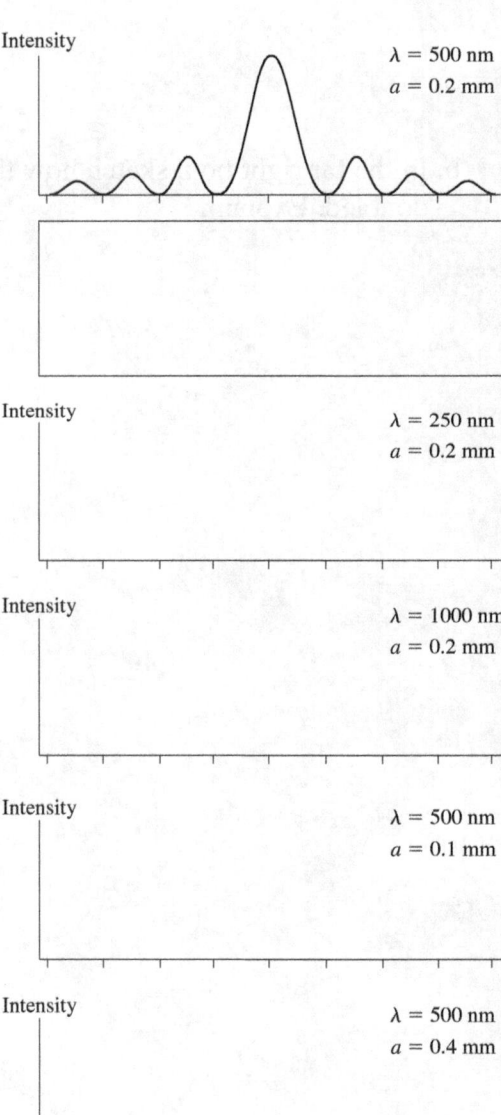

17.6 Circular-Aperture Diffraction

16. This is the light intensity on a viewing screen behind a circular aperture.

a. In the middle box, sketch how the pattern would appear if the wavelength of the light is doubled. Explain.

b. In the far right box, sketch how the pattern would appear if the diameter of the aperture is doubled. Explain.

18 Ray Optics

Note: Please use a ruler or straight edge for drawing light rays.

18.1 The Ray Model of Light

1. a. Draw four or five rays from the object that allow A to see the object.
   b. Draw four or five rays from the object that allow B to see the object.

   c. Describe the situations seen by A and B if a piece of cardboard is lowered at point C.

2. a. Draw three or four rays from object 1 that allow A to see object 1.
   b. Draw three or four rays from object 2 that allow B to see object 2.
   c. What, if anything, happens to the light where the rays cross in the center of the picture?

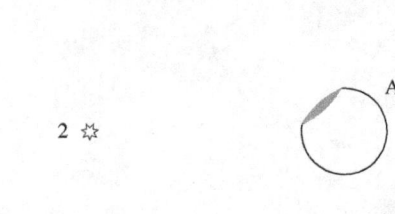

3. A point source of light illuminates a slit in an opaque barrier.

   a. On the screen, sketch the pattern of light that you expect to see. Let the white of the paper represent light areas; shade dark areas. Mark any relevant dimensions. **Note:** This slit is too wide to observe diffraction.

   b. What will happen to the pattern of light on the screen if the slit width is reduced to 0.5 cm?

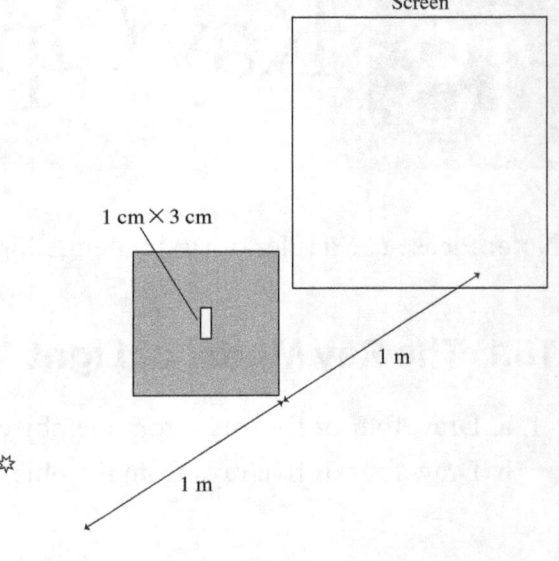

4. In each situation below, light passes through a 1-cm-diameter hole and is viewed on a screen. For each, sketch the pattern of light that you expect to see on the screen. Let the white of the paper represent light areas; shade dark areas.

   a.　　　　　　　　b.　　　　　　　　c.

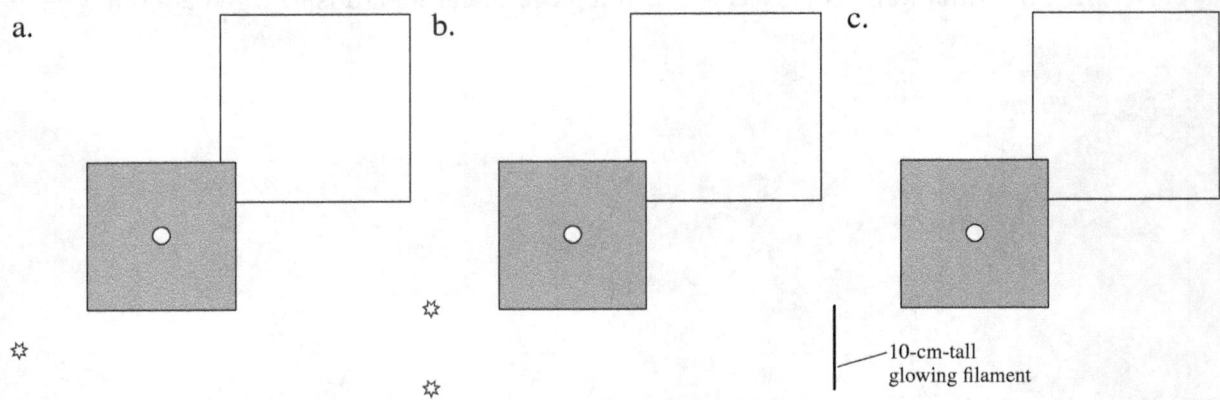

5. Light from an L-shaped bulb passes through a pinhole. On the screen, sketch the pattern of light that you expect to see.

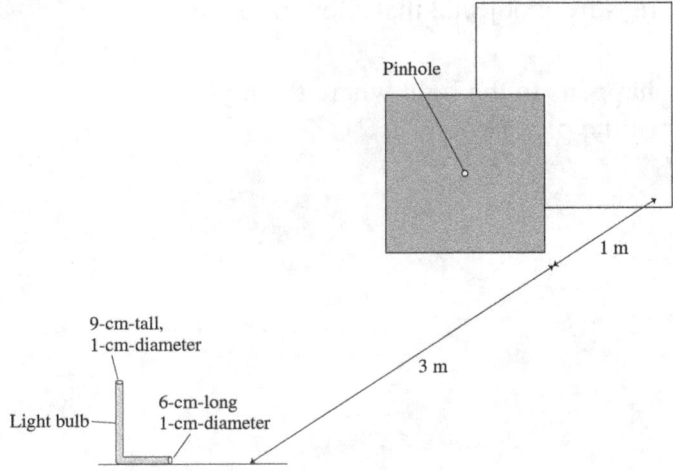

18.2 Reflection

6. a. Use a straight edge to draw five rays from the object that pass through points A to E after reflecting from the mirror. Make use of the grid to do this accurately.

   b. Extend the reflected rays behind the mirror.

   c. Show and label the image point.

7. a. Draw *one* ray from the object that enters the eye after reflecting from the mirror.

   b. Is one ray sufficient to tell your eye/brain where the image is located?

   c. Use a different color pen or pencil to draw two more rays that enter the eye after reflecting. Then use the three rays to locate (and label) the image point.

   d. Do any of the rays that enter the eye actually pass through the image point?

8. You are looking at the image of a pencil in a mirror.
   a. What happens to the image you see if the top half of the mirror, down to the midpoint, is covered with a piece of cardboard? Explain.

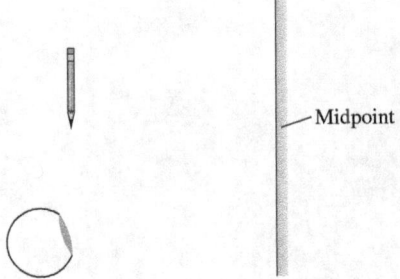

Midpoint

   b. What happens to the image you see if the bottom half of the mirror is covered with a piece of cardboard? Explain.

9. Two parallel mirrors face each other.

    a. Draw rays from the object that reflect from mirror 1 at the two dots. Use a straight edge and the grid to draw the reflected rays accurately.

    b. Extend the reflected rays backward to locate the object's image in mirror 1. Use a dot to indicate the image point, and label it Image 1.

    c. The rays that reflect from mirror 1 then reflect from mirror 2. Use a straight edge and the grid to accurately draw the rays reflecting from mirror 2.

    d. Extend the reflected rays backward. The point from which they appear to originate is the image, in mirror 2, of image 1. Use a dot to indicate this point, and label it Image 2.

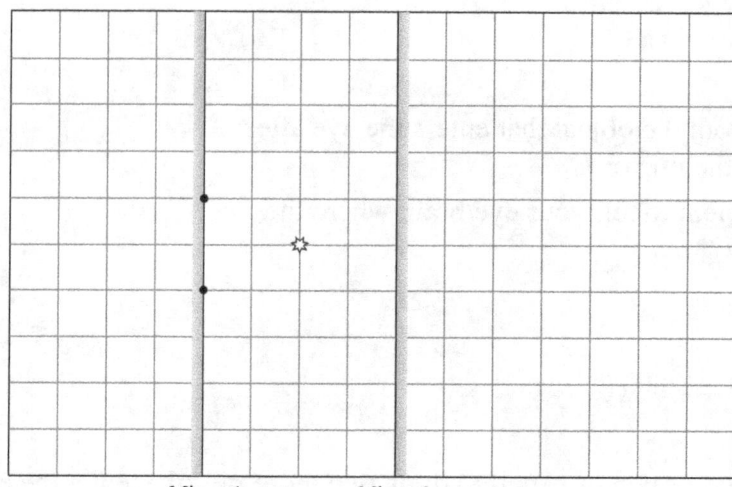

        Mirror 1   Mirror 2

18.3 Refraction

10. Use a straight edge to complete the trajectories of these three rays through material 2 and back into material 1. Assume $n_2 < n_1$.

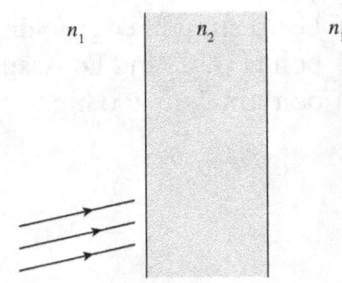

11. The figure shows six conceivable trajectories of light rays leaving an object. Which, if any, of these trajectories are impossible? For each that is possible, what are the requirements of the index of refraction n_2?

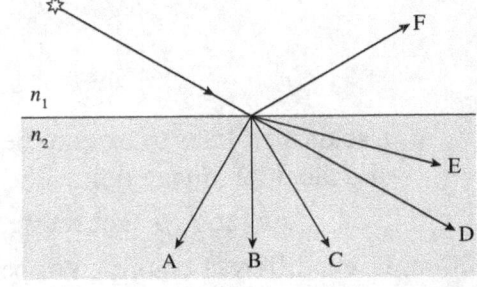

Impossible _____

Requires $n_2 > n_1$ _____

Requires $n_2 = n_1$ _____

Requires $n_2 < n_1$ _____

Possible for any n_2 _____

12. Complete the ray trajectories through the two prisms shown below.

a.

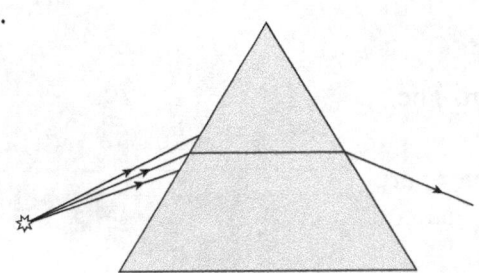

b.

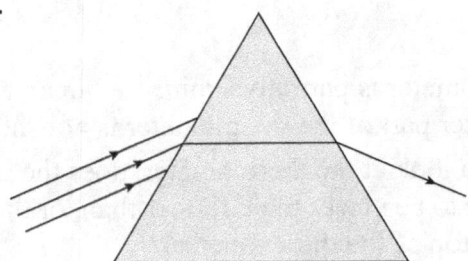

13. Draw the trajectories of seven rays that leave the object heading toward the seven dots on the boundary. Assume $n_2 < n_1$ and $\theta_c = 47°$.

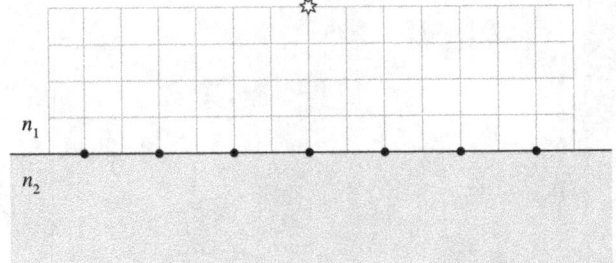

18.4 Image Formation by Refraction

14. a. Use a straight edge to draw rays that leave the object and refract after passing through points B, C, and D. Assume $n_2 > n_1$. The refraction at B and D should be the same size—don't make it too big.

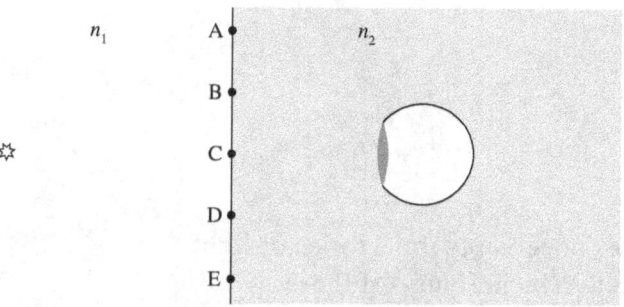

   b. Use dashed lines to extend the three refracted rays backward into medium 1. Then locate and label the image point.

   c. Now draw the rays that refract at A and E.

   d. Use a different color pen or pencil to draw three rays from the object that enter the eye.

   e. Does the distance to the object *appear* to be greater than, less than, or the same as the true distance? Explain.

15. A thermometer is partially submerged in an aquarium. The underwater part of the thermometer is not shown.

   a. As you look at the thermometer, does the underwater part appear to be closer than, farther than, or the same distance as the top of the thermometer?

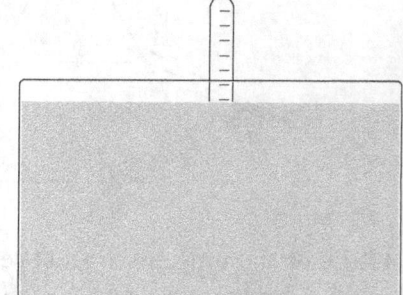

   b. Complete the drawing by drawing the bottom of the thermometer as you think it would look.

18.5 Thin Lenses: Ray Tracing

16. a. Continue these rays through the lens and out the other side.

    b. Is the point where the rays converge the same as the focal point of the lens? Or different? Explain.

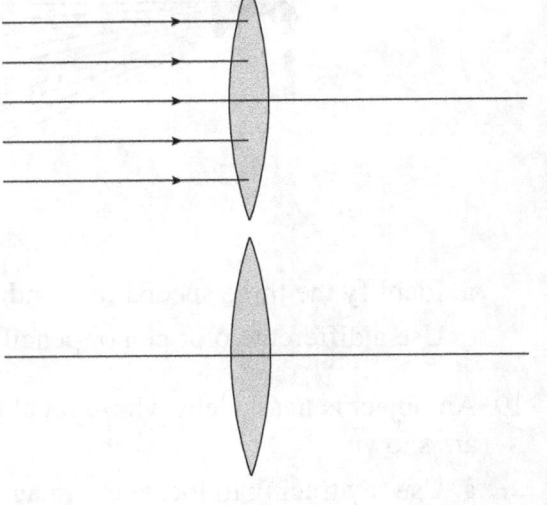

    c. Place a point source of light at the place where the rays converged in part b. Draw several rays heading left, toward the lens. Continue the rays through the lens and out the other side.

    d. Do these rays converge? If so, where?

17. The top two figures show test data for a lens. The third figure shows a point source near this lens and four rays heading toward the lens.

    a. For which of these rays do you know, from the test data, its direction after passing through the lens?

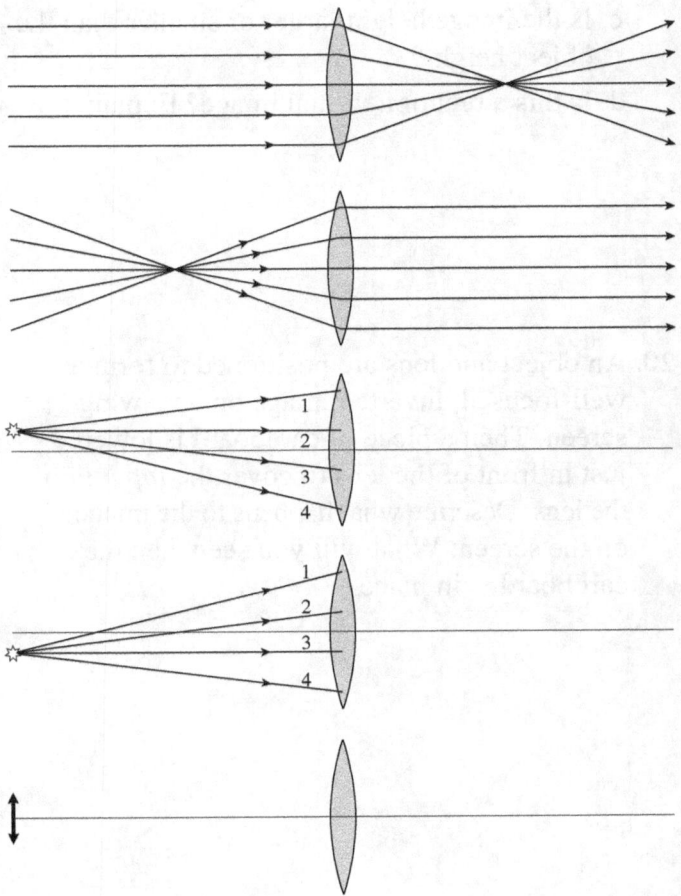

    b. Draw the rays you identified in part a as they pass through the lens and out the other side.

    c. Use a different color pen or pencil to draw the trajectories of the other rays.

    d. Label the image point. What kind of image is this?

    e. The fourth figure shows a second point source. Use ray tracing to locate its image point.

    f. The fifth figure shows an extended object. Have you learned enough to locate its image? If so, draw it.

18. An object is near a lens whose focal points are marked with dots.

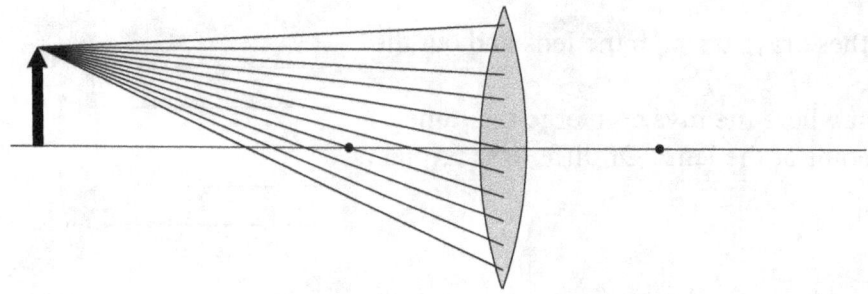

    a. Identify the three special rays and use a straight edge to continue them through the lens.

    b. Use a different color pen or pencil to draw the trajectories of the other rays.

19. An object is near a lens whose focal points are shown.

    a. Use ray tracing to locate the image of this object.

    b. Is the image upright or inverted?_____

    c. Is the image height larger or smaller than the object height?_____

    d. Is this a real or a virtual image? Explain how you can tell.

20. An object and lens are positioned to form a well-focused, inverted image on a viewing screen. Then a piece of cardboard is lowered just in front of the lens to cover the *top half* of the lens. Describe what happens to the image on the screen. What will you see when the cardboard is in place?

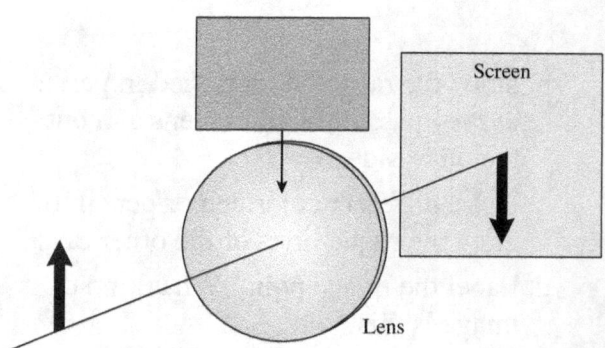

18.6 Image Formation with Spherical Mirrors

21. Two spherical mirrors are shown. The center of each sphere is marked with an open circle. For each:
    i. Use a straight edge to draw the normal to the surface at the seven dots on the boundary.
    ii. Draw the trajectories of seven rays that leave the object, strike the mirror surface at the dots, and then reflect, obeying the law of reflection.
    iii. Trace the reflected rays either forward to a point where they converge or backward to a point from which they diverge.

a.   b.

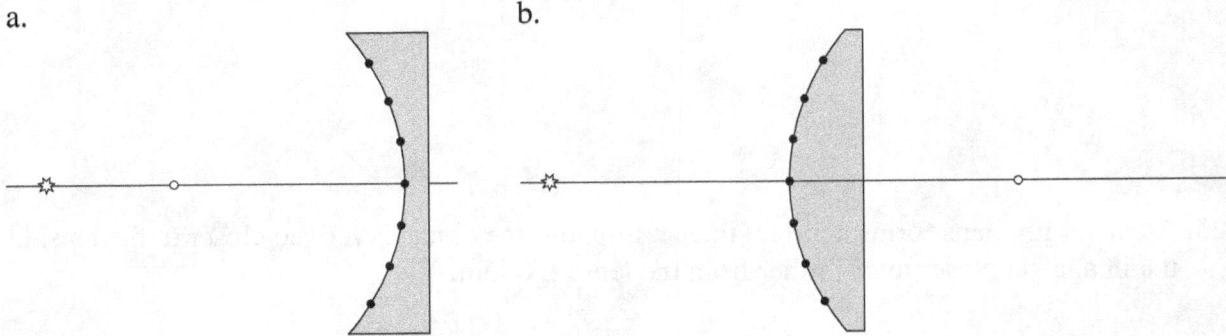

22. An object is placed near a spherical mirror whose focal point is marked.

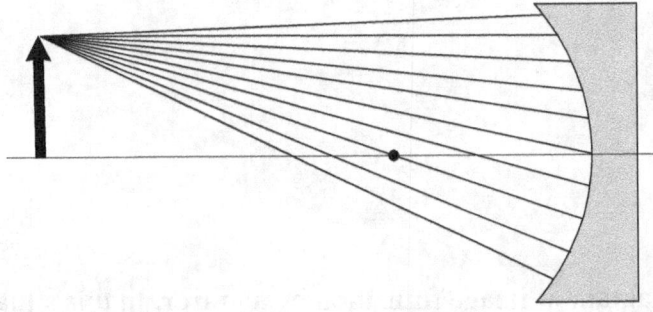

    a. Identify the three special rays and show their reflections.
    b. Use a different color pen or pencil to draw the trajectories of the other rays.

23. A 3.0-cm-high object is placed 10.0 cm in front of a convex diverging mirror with a focal length of −4.0 cm. Use ray tracing to determine the location of the image, the orientation of the image, and the height of the image.
    • Locate the mirror on the optical axis shown. Show its focal point with a dot.
    • Represent the object with an upright arrow at distance 10.0 cm from the mirror.
    • Draw the appropriate three "special rays" to locate the image.

18.7 The Thin-Lens Equation

24. A converging lens forms a real image. Suppose the object is moved farther from the lens. Does the image get closer to or farther from the lens? Explain.

25. A converging lens forms a virtual image. Suppose the object is moved closer to the lens. Does the image get closer to or farther from the lens? Explain.

26. The figure is a ray diagram of image formation by a mirror. In this situation,

    Is s positive, negative, or zero? _____

    Is s' positive, negative, or zero? _____

    Is f positive, negative, or zero? _____

    Is m positive, negative, or zero? _____

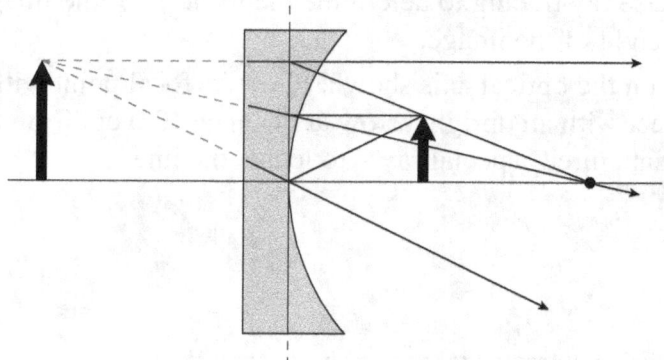

19 Optical Instruments

19.1 The Camera

19.2 The Human Eye

1. A photographer focuses his camera on an object. Suppose the object moves closer to the camera. To refocus, should the camera lens move closer to or farther from the detector? Explain.

2. Two lost students wish to start a fire to keep warm while they wait to be rescued. One student is hyperopic, the other myopic. Which, if either, could use his glasses to focus the sun's rays to an intense bright point of light? Explain.

3. Suppose you wanted special glasses designed to let you see underwater, without a face mask. Should the glasses use a converging or diverging lens? Explain.

4. Equip each eyeball below with an appropriate eyeglass lens that will produce a well-focused image on the retina. To do so, first draw the lens in front of the eye, then redraw the two rays from the point at which they enter the lens until they form an image.

a.

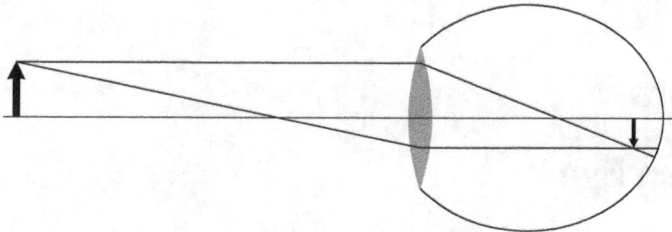

b. Is this eye hyperopic or myopic? _____ Explain.

c.

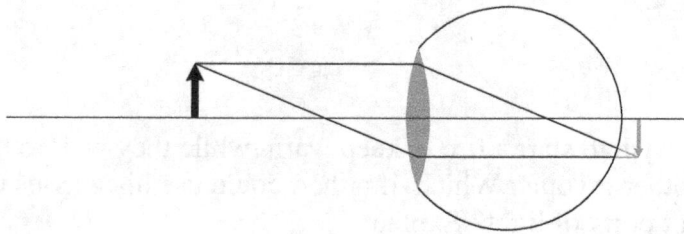

d. Is this eye hyperopic or myopic? _____ Explain.

19.3 The Magnifier

5. An eye views objects A and B.
   a. Which object has the larger size? Explain.

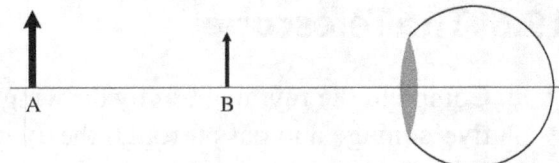

   b. Which object has the larger angular size? Explain.

6. The angular magnification of a lens is not sufficient. To double the angular magnification, do you want a lens with twice the focal length or half the focal length? Explain.

7. On the left, an eye observes an object at the eye's near point of 25 cm. This is the closest the object can be and still be seen clearly. On the right, the eye views the same object through a magnifying lens. The object's physical distance from the eye is now much less than 25 cm.

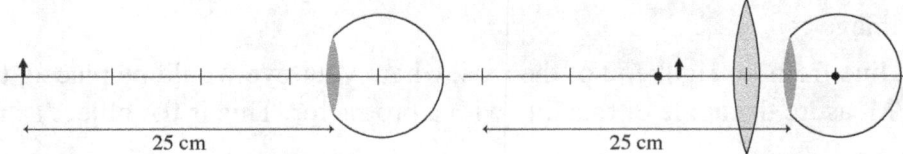

   a. Using a straight edge, draw a line on the left figure to indicate the angular size of the object when viewed by the unaided eye at the eye's near point.
   b. Use ray tracing with a straight edge to show that the image in the right figure is a virtual image ≈ 25 cm from the eye.
   c. Draw a line (or label one of your ray-tracing lines) to indicate the angular size of the image seen through the lens.
   d. Using a ruler to make measurements, determine the angular magnification of the lens.

   e. It's sometimes said that a magnifying glass makes an object "appear closer." Is that what is happening here? Explain.

19.4 The Microscope

19.5 The Telescope

8. a. Complete the ray diagram by drawing two special rays that start from the tip of the objective's image and pass through the eyepiece. Show that the two rays are parallel on the right side of the eyepiece. (Because these rays are parallel, it is not possible to draw the final virtual image on your diagram.)

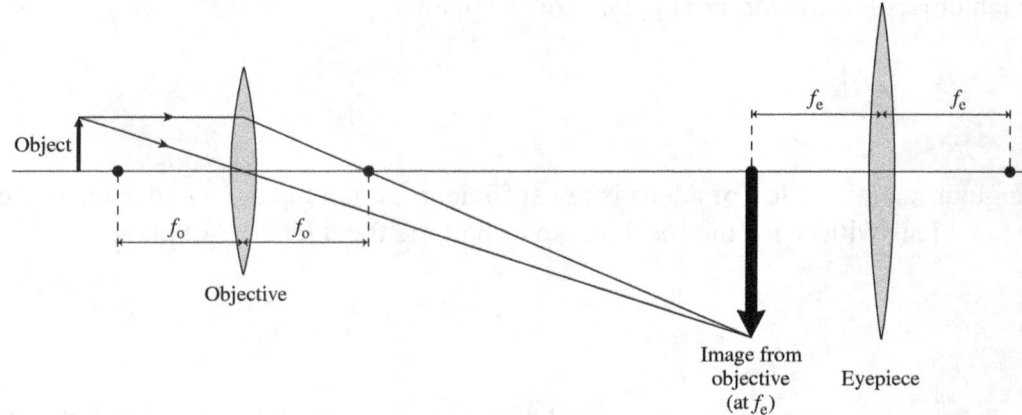

b. On the diagram above, indicate the angle subtended by the final image. This is the image's angular size. Measure this angle with a protractor.

θ final image = _____

c. Draw a line from far right end of the axis, where your eye would be placed, to the tip of the object. Measure the angle of this line with a protractor. This is the object's angular size.

θ object = _____

d. What is the angular magnification of the two-lens system?

9. Rank in order, from largest to smallest, the magnifications M_1 to M_4 of these telescopes.

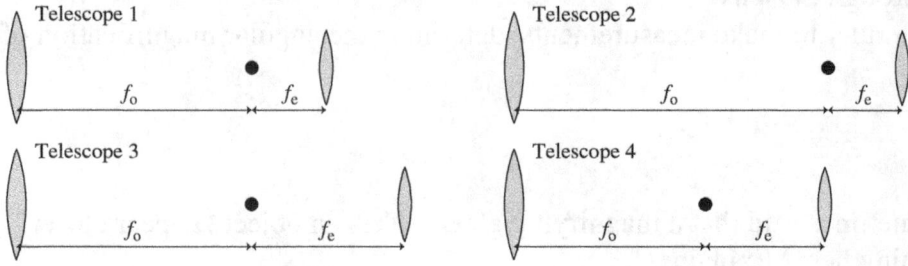

Order:

Explanation:

19.6 Color and Dispersion

10. A beam of white light from a flashlight passes through a red piece of plastic.

    a. What is the color of the light that emerges from the plastic? _____

    b. Is the emerging light as intense as, more intense than, or less intense than the white light? Explain.

    c. The light then passes through a blue piece of plastic. Describe the color and intensity of the light that emerges.

11. Suppose you looked at the sky on a clear day through pieces of red and blue plastic oriented as shown. Describe the color and brightness of the light coming through sections 1, 2, and 3.

    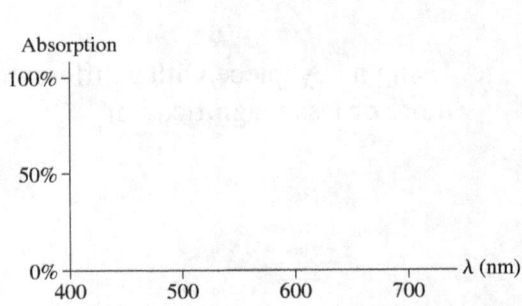

    Section 1:

    Section 2:

    Section 3:

12. Sketch a plausible absorption spectrum for a patch of bright red paint.

    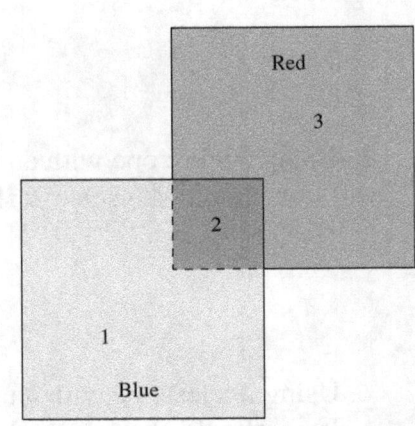

19.7 Resolution of Optical Instruments

13. A diffraction-limited lens can focus light to a 10-μm-diameter spot on a screen. Do the following actions make the spot diameter larger, smaller, or leave it unchanged?

    a. Decreasing the wavelength of light: _____

    b. Decreasing the lens diameter: _____

    c. Decreasing the focal length: _____

    d. Decreasing the lens-to-screen distance: _____

14. An astronomer is trying to observe two distant stars. The stars are marginally resolved when she looks at them through a filter that passes green light near 550 nm. Which of the following actions would improve the resolution? Assume that the resolution is not limited by the atmosphere.

    a. Changing the filter to a different wavelength? If so, should she use a shorter or a longer wavelength?

    b. Using a telescope with an objective lens of the same diameter but different focal length? If so, should she select a shorter or a longer focal length?

    c. Using a telescope with an objective lens of the same focal length but a different diameter? If so, should she select a larger or a smaller diameter?

    d. Using an eyepiece with a different magnification? If so, should she select an eyepiece with more or less magnification?

20 Electric Fields and Forces

20.1 Charges and Forces

1. Two lightweight balls hang straight down when both are neutral. They are close enough together to interact, but not close enough to touch. Draw pictures showing how the balls hang if: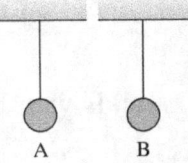

   a. Both are touched with a plastic rod that was rubbed with wool.

   b. The two charged balls of part a are moved farther apart.

   c. Ball A is touched by a plastic rod that was rubbed with wool and ball B is touched by a glass rod that was rubbed with silk.

   d. Both are charged by a plastic rod, but ball A is charged more than ball B.

   e. Ball A is charged by a plastic rod. Ball B is neutral.

2. After combing your hair briskly, the comb will pick up small pieces of paper.

a. Is the comb charged? Explain.

b. How can you be sure that it isn't the paper that is charged? Propose an experiment to test this.

c. Is your hair charged after being combed? What evidence do you have for your answer?

d. What kind of charge—positive or negative—is the comb likely to have? Why?

e. How could you test your answer to part d?

3. A negatively charged electroscope has separated leaves.

a. Suppose you bring a negatively charged rod close to the top of the electroscope, but not touching. How will the leaves respond? Use both charge diagrams and words to explain.

b. How will the leaves respond if you bring a positively charged rod close to the top of the electroscope, but not touching? Use both charge diagrams and words to explain.

a.

b.

4. Four lightweight balls A, B, C, and D are suspended by threads. Ball A has been touched by a plastic rod that was rubbed with wool. When the balls are brought close together, without touching, the following observations are made:

   • Balls B, C, and D are attracted to ball A.
   • Balls B and D have no effect on each other.
   • Ball B is attracted to ball C.

   What are the charge states (positive, negative, or neutral) of balls A, B, C, and D? Explain.

5. a. Metal sphere A is initially neutral. A positively charged rod is brought near, but not touching. Is A now positive, negative, or neutral? Explain.

   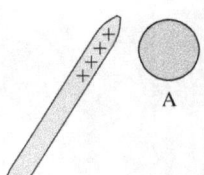

   b. Metal spheres A and B are initially neutral and are touching. A positively charged rod is brought near A, but not touching. Is A now positive, negative, or neutral? Explain.

   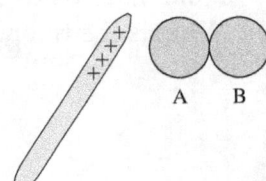

   c. Metal sphere A is initially neutral. It is connected by a metal wire to the ground. A positively charged rod is brought near, but not touching. Is A now positive, negative, or neutral? Explain.

   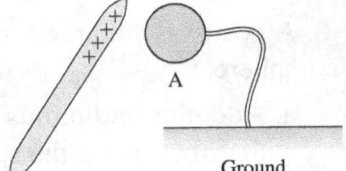

20.2 Charges, Atoms, and Molecules

6. Two oppositely charged metal spheres have equal quantities of charge. They are brought into contact with a neutral metal rod.

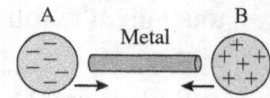

a. What is the final charge state of each sphere and of the rod?

b. Use both charge diagrams and words to explain how this charge state is reached.

7. Metal sphere A has 4 units of negative charge and metal sphere B has 2 units of positive charge. The two spheres are brought into contact. What is the final charge state of each sphere? Explain.

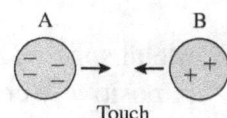

8. A positively charged rod is held near, but not touching, a neutral metal sphere.

a. Add plus and minus signs to the figure to show the charge distribution on the sphere.

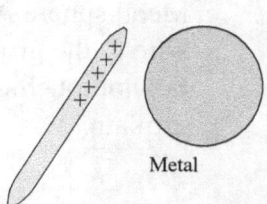

b. Does the sphere experience a net force? If so, in which direction? Explain.

20.3 Coulomb's Law

9. For each pair of charges, draw a force vector *on each charge* to show the electric force acting on that charge. The length of each vector should be proportional to the magnitude of the force. Each + and − symbol represents the same quantity of charge.

a.

b.

c.

d.

10. For each group of charges, use a **black** pen or pencil to draw each force acting on the gray positive charge. Then use a **red** pen or pencil to show the net force on the gray charge. Label $\vec{F}_{net}$.

a.

b.

c.

11. Can you assign charges (positive or negative) so that these forces are correct? If so, show the charges on the figure. (There may be more than one correct response.) If not, why not?

a.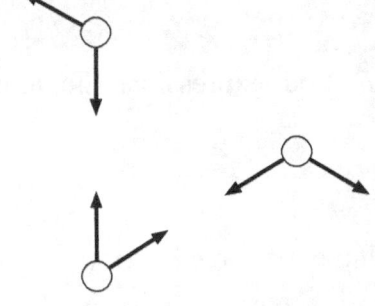

b.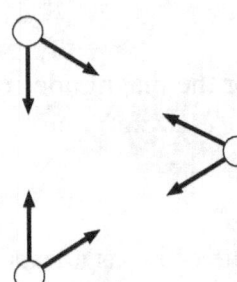

c.

d.

12. Draw a + on the figure below to show the position or positions where a proton would experience no net force.

13. Draw a − on the figure below to show the position or positions where an electron would experience no net force.

14. The gray positive charge experiences a net force due to two other charges: the +1 charge that is seen and a +4 charge that is not seen. Add the +4 charge to the figure at the correct position.

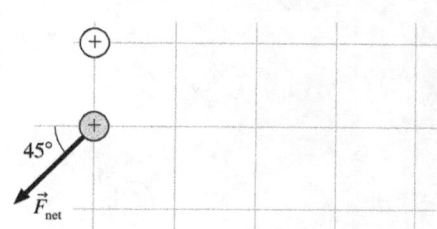

15. Positive charges $4q$ and q are distance L apart. Let them be on the
PSA x-axis with $4q$ at the origin.
20.1

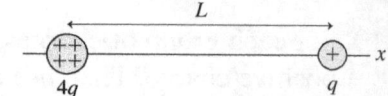

   a. Suppose a proton were placed on the x-axis to the *right* of q. Is it *possible* that the net electric force on the proton is zero? Explain.

   b. On the figure, draw a proton at an arbitrary point on the x-axis between $4q$ and q. Label its distance from $4q$ as r. Draw two force vectors and label them $\vec{F}_{4q}$ and $\vec{F}_q$ to show the two forces on this proton. Is it *possible* that, for the proper choice of r, the net electric force on the proton is zero? Explain.

   c. Write expressions for the magnitudes of forces $\vec{F}_{4q}$ and $\vec{F}_q$ Your expressions should be in terms of K, q, L, and r.

   $$F_{4q} = \underline{\hspace{3in}} \qquad F_q = \underline{\hspace{2in}}$$

   d. Find the specific value of r—as a fraction of L—at which the net force is zero.

20.4 The Concept of the Electric Field

16. At points 1 to 4, draw an electric field vector with the proper direction and with a length proportional to the electric field strength at that point.

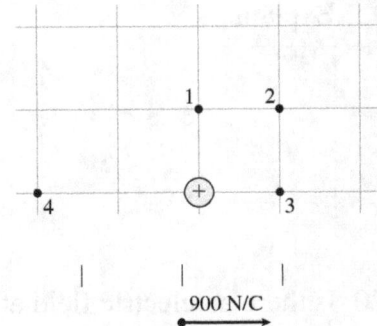

17. a. The electric field of a point charge is shown at *one* point in space.
    Can you tell if the charge is + or −? If not, why not?

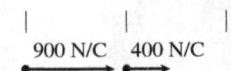

   b. Here the electric field of a point charge is shown at two positions in space.
   Now can you tell if the charge is + or −? Explain.

   c. Can you determine the location of the charge? If so, draw it on the figure. If not, why not?

18. The electric field strength at a point in space near a point charge is 1000 N/C.
    a. What will be the field strength if the quantity of charge is halved? Explain.

    b. What will be the field strength if the distance to the point charge is halved? The quantity of charge is the original amount, not the value of part a. Explain.

19. Is the electric field strength at point A larger than, smaller than, or the same as the electric field strength at point B? Explain.

20. Is there an electric field at the position of the dot? If so, draw the electric field vector on the figure. If not, what would you need to do to create an electric field at this point?

20.5 The Electric Field of Multiple Charges

21. At each of the dots, use a **black** pen or pencil to draw and label the electric fields $\vec{E}_1$ and $\vec{E}_2$ due to the two point charges. Make sure that the *relative* lengths of your vectors indicate the strength of each electric field. Then use a **red** pen or pencil to draw and label the net electric field $\vec{E}_{net}$.

a.

b.

22. For each of the figures, use dots to mark any point or points (other than infinity) where $\vec{E} = \vec{0}$.

a.

b.

23. The figure shows the electric field lines in a region of space. Draw the electric field vectors at the three dots.

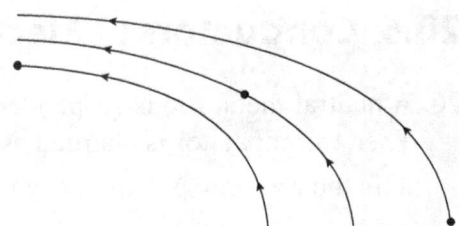

24. Rank in order, from largest to smallest, the electric field strengths E_1 to E_5 at each of these points.

   Order:

   Explanation:

25. A parallel-plate capacitor is constructed of two square plates, size $L \times L$, separated by distance d. The plates are given charge $\pm Q$. What is the ratio E_f/E_i of the final electric field strength E_f to the initial electric field strength E_i if:

   a. Q is doubled?

   b. L is doubled?

   c. d is doubled?

20.6 Conductors in Electric Fields

26. A neutral metal rod is suspended in the center of a parallel-plate capacitor. Then the capacitor is charged as shown.

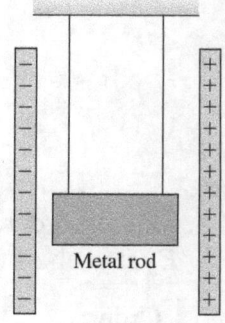

   a. Is the rod now positive, negative, or neutral? Explain.

   b. Is the rod polarized? If so, draw plus and minus signs on the figure to show the charge distribution. If not, why not?

   c. Does the rod swing toward one of the plates, or does it remain in the center? If it swings, which way? Explain.

20.7 Forces and Torques in Electric Fields

27. Positively and negatively charged particles, with equal masses and equal quantities of charge, are shot into a capacitor in the directions shown.

   a. Use solid lines to draw their trajectories on the figure if their initial velocities are fast.

   b. Use dotted lines to draw their trajectories on the figure if their initial velocities are slow.

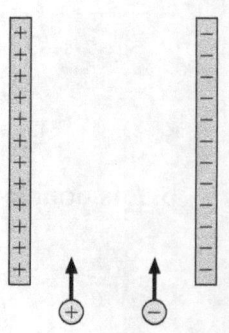

28. An electron is launched from the positive plate at a 45° angle. It does not have sufficient speed to make it to the negative plate. Draw its trajectory on the figure.

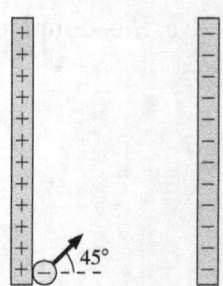

29. Three charges are placed at the corners of a triangle. The ++charge
    has twice the quantity of charge of the two − charges; the net charge is
    zero.

    a. Draw the force vectors on each of the charges.

    b. Is the triangle in equilibrium? _____ If not, draw the
       equilibrium orientation directly beneath the triangle that is shown.

    c. Once in the equilibrium orientation, will the triangle move to the
       right, move to the left, rotate steadily, or be at rest? Explain.

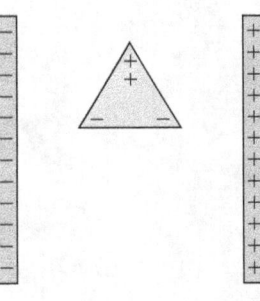

You Write the Problem!

Exercises 30–32: You are given the equation that is used to solve a problem. For each of these:

    a. Write a *realistic* physics problem for which this is the correct equation. Look at worked
       examples and end-of-chapter problems in the textbook to see what realistic physics
       problems are like.

    b. Finish the solution of the problem.

30. $\dfrac{(9.0 \times 10^9 \ \text{N} \cdot \text{m}^2/\text{C}^2) \times q^2}{(0.015 \ \text{m})^2} = 0.020 \ \text{N}$

31. $\dfrac{(9.0 \times 10^9 \ \text{N} \cdot \text{m}^2/\text{C}^2) \times N \times (1.6 \times 10^{-19} \ \text{C})}{(1.0 \times 10^{-6} \ \text{m})^2} = 1.5 \times 10^6 \ \text{N/C}$

32. $1.5 \times 10^6 \ \text{N/C} = \dfrac{Q}{(8.85 \times 10^{-12} \ \text{C}^2/\text{N} \cdot \text{m}^2) \, \pi \, (0.0125 \ \text{m})^2}$

21 Electric Potential

21.1 Electric Potential Energy and Electric Potential

1. A force does 2 μJ of work to push charged particle A toward a set of fixed source charges. Charged particle B has twice the charge of A. How much work must the force do to push B through the same displacement? Explain.

2. A 1 nC charged particle is pushed toward a set of fixed source charges, as shown. In the process, the particle gains 1 μJ of electric potential energy.

   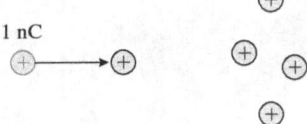

   Source charges

   a. How much work was done to push the particle through this displacement? Explain.

   b. Through what potential difference did the particle move?

3. Charged particle A is placed at a point in space where the electric potential is V. Its electric potential energy at that point is U_A. Particle A is removed and replaced by charged particle B, whose potential energy at the same point is U_B. If the charge of B is three times the charge of A, what is the ratio U_B/U_A? Explain.

4. Which point, A or B, has the higher electric potential? Why?

   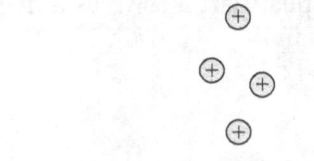

   Source charges

21.2 Sources of Electric Potential

21.3 Electric Potential and Conservation of Energy

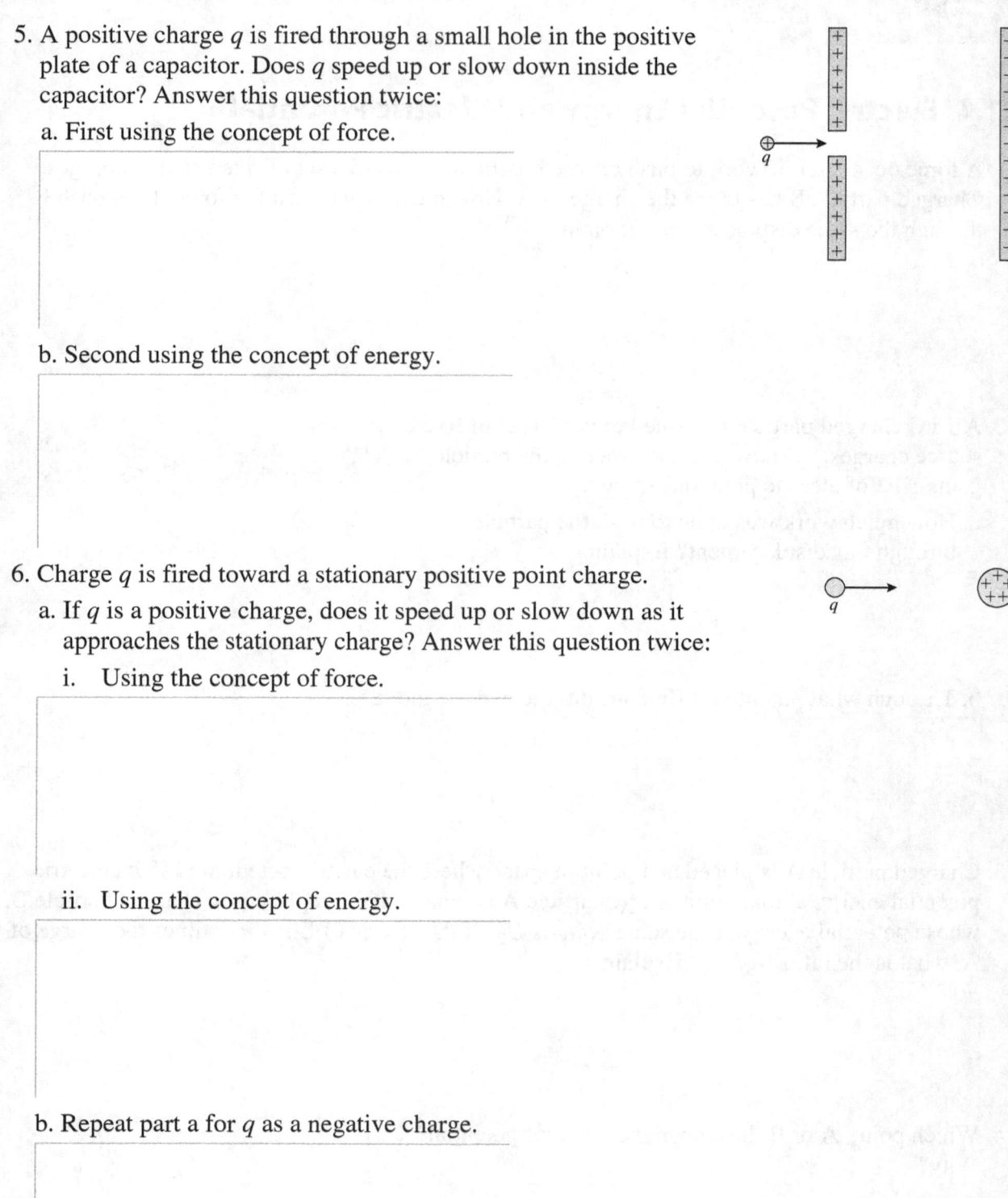

5. A positive charge q is fired through a small hole in the positive plate of a capacitor. Does q speed up or slow down inside the capacitor? Answer this question twice:

   a. First using the concept of force.

   b. Second using the concept of energy.

6. Charge q is fired toward a stationary positive point charge.

   a. If q is a positive charge, does it speed up or slow down as it approaches the stationary charge? Answer this question twice:

   i.   Using the concept of force.

   ii. Using the concept of energy.

   b. Repeat part a for q as a negative charge.

21.4 Calculating the Electric Potential

7. Rank in order, from largest to smallest, the electric potentials V_1 to V_5 at points 1 to 5.

   Order:

   Explanation:

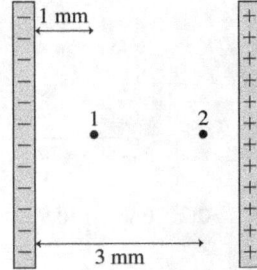

8. The figure shows two points inside a capacitor. Let $V = 0$ V at the negative plate.

   a. What is the ratio V_2/V_1 of the electric potentials at these two points? Explain.

   b. What is the ratio E_2/E_1 of the electric field strengths at these two points? Explain.

9. A capacitor with plates separated by distance d is charged to a potential difference ΔV_C. All wires and batteries are disconnected, and then the two plates are pulled apart (with insulated handles) to a new separation of distance $2d$.

   a. Does the capacitor charge Q change as the separation increases? If so, by what factor? If not, why not?

   b. Does the electric field strength E change as the separation increases? If so, by what factor? If not, why not?

   c. Does the potential difference ΔV_C change as the separation increases? If so, by what factor? If not, why not?

10. Each figure shows a contour map on the left and a set of graph axes on the right. Draw a graph of V versus x. Your graph should be a straight line or a smooth curve.

a.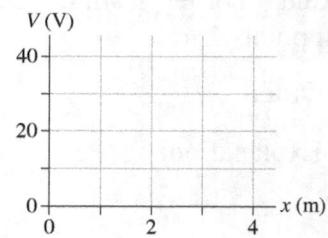

b.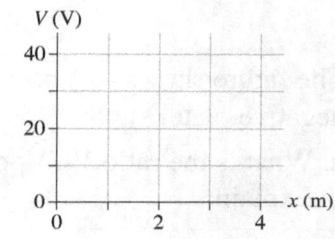

c.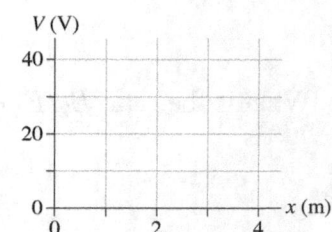

11. Each figure shows a V-versus-x graph on the left and an x-axis on the right. Assume that the potential varies with x but not with y. Draw a contour map of the electric potential. There should be a uniform potential difference between equipotential lines, and each equipotential line should be labeled.

a.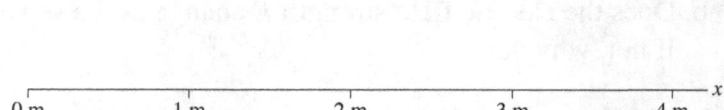

b.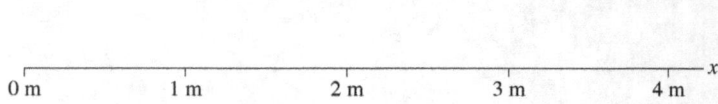

12. Rank in order, from largest to smallest, the electric potentials V_1 to V_5 at points 1 to 5.

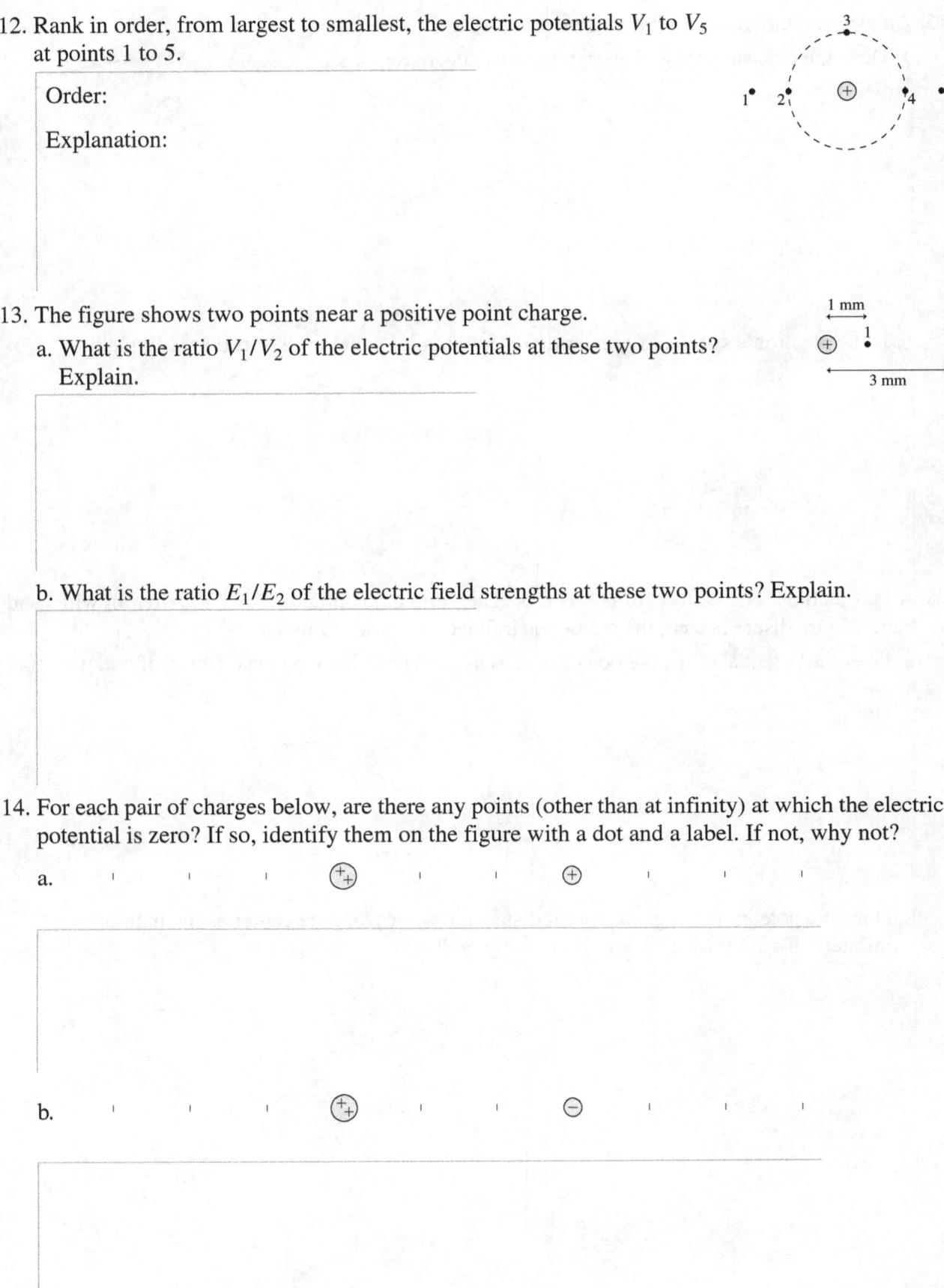

> Order:
>
> Explanation:

13. The figure shows two points near a positive point charge.

    a. What is the ratio V_1/V_2 of the electric potentials at these two points? Explain.

    b. What is the ratio E_1/E_2 of the electric field strengths at these two points? Explain.

14. For each pair of charges below, are there any points (other than at infinity) at which the electric potential is zero? If so, identify them on the figure with a dot and a label. If not, why not?

    a.

    b.

15. An electron moves along the trajectory from i to f.

    a. Does the electric potential energy increase, decrease, or stay the same? Explain.

    b. Is the electron's speed at f greater than, less than, or equal to its speed at i? Explain.

16. An inflatable metal balloon of radius R is charged to a potential of 1000 V. After all wires and batteries are disconnected, the balloon is inflated to a new radius $2R$.

    a. Does the potential of the balloon change as it is inflated? If so, by what factor? If not, why not?

    b. Does the potential change at a point distance $r = 4R$ from the center as the balloon is inflated? If so, by what factor? If not, why not?

17. A small charged sphere of radius R_1, mass m_1, and positive charge q_1 is shot head on with
    PSA speed v_1 from a long distance away toward a second small sphere having radius R_2, mass m_2,
    21.1 and positive charge q_2. The second sphere is held in a fixed location and cannot move. The
    spheres repel each other, so sphere 1 will slow as it approaches sphere 2. If v_1 is small, sphere
    1 will reach a closest point, reverse direction, and be pushed away by sphere 2. If v_1 is large,
    sphere 1 will crash into sphere 2. For what speed v_1 does sphere 1 just barely touch sphere 2 as
    it reverses direction?

    a. Begin by drawing a before-and-after visual overview. Initially, the spheres are far apart and
       sphere 1 is heading toward sphere 2 with speed v_1. The problem ends with the spheres
       touching. What is speed of sphere 1 at this instant? How far apart are the centers of the
       spheres at this instant? Label the before and after pictures with complete information—all in
       symbolic form.

    b. Energy is conserved, so we can use Problem-Solving Approach 21.1. But first we have to
       identify the "moving charge" q and the "source charge" that creates the potential.

       Which is the moving charge? _____ Which is the source charge? _____

    c. We're told the charges start "a long distance away" from each other. Based on this
       statement, what value can you assign to V_i, the potential of the source charge at the initial
       position of the moving charge? Explain.

    d. Now write an expression in terms of the symbols defined above (and any constants that are
       needed) for the initial energy $K_i + qV_i$.

       $K_i + qV_i = $ _____

    e. Referring to information on your visual overview, write an expression for the final energy.

       $K_f + qV_f = $ _____

    f. Energy is conserved, so finish the problem by solving for v_1.

21.5 Connecting Potential and Field

18. For each contour map:
    i.   Estimate the electric fields $\vec{E}_a$ and $\vec{E}_b$ at points a and b. Don't forget that $\vec{E}$ is a vector. Show how you made your estimate.
    ii. Draw electric field vectors on top of the contour map.

a.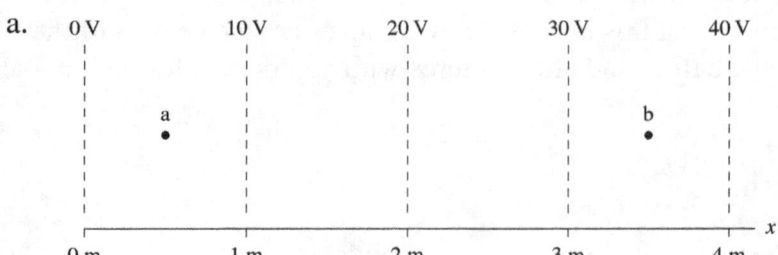

$\vec{E}_a = \underline{\hspace{4cm}}$

$\vec{E}_b = \underline{\hspace{4cm}}$

b.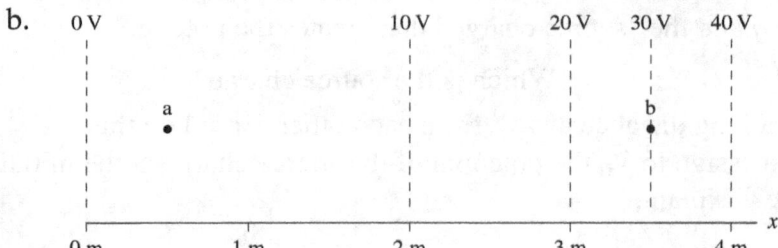

$\vec{E}_a = \underline{\hspace{4cm}}$

$\vec{E}_b = \underline{\hspace{4cm}}$

19. Draw the electric field vectors at the dots on this contour map. The length of each vector should be proportional to the field strength at that point.

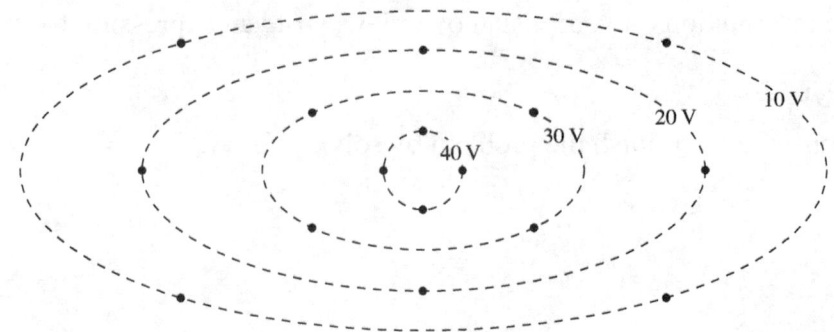

20. Two metal spheres are connected by a metal wire that has a switch in the middle. Initially, the switch is open. Sphere 1, with the larger radius, is given a positive charge. Sphere 2, with the smaller radius, is neutral. Then the switch is closed. Afterward, sphere 1 has charge Q_1, is at potential V_1, and the electric field strength at its surface is E_1. The values for sphere 2 are Q_2, V_2, and E_2.

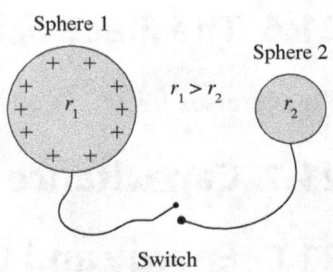

a. Is V_1 larger than, smaller than, or equal to V_2? Explain.

b. Is Q_1 larger than, smaller than, or equal to Q_2? Explain.

c. Is E_1 larger than, smaller than, or equal to E_2? Explain.

21. The figure shows a hollow metal sphere. A negatively charged rod touches the top of the sphere, transferring charge to the sphere. Then the rod is removed.

a. Show on the figure the equilibrium distribution of charge.

b. Draw the electric field diagram.

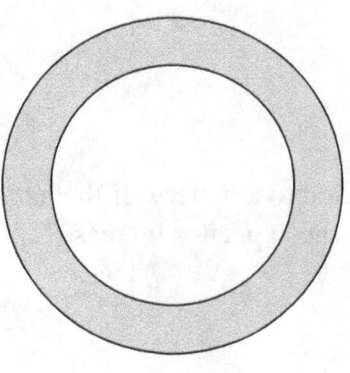

21.6 The Electrocardiogram

No exercises for this section.

21.7 Capacitance and Capacitors

21.8 Energy and Capacitors

22. Rank in order, from largest to smallest, the potential differences $(\Delta V_C)_1$ to $(\Delta V_C)_4$ of these four capacitors.

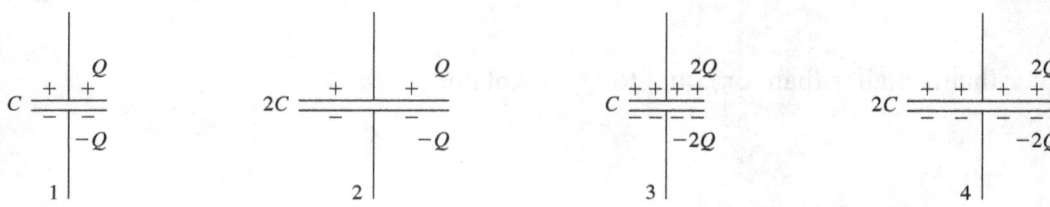

Order:

Explanation:

23. A parallel-plate capacitor has capacitance C. Suppose all three dimensions of the capacitor are doubled—that is, increased by a factor of 2. By what factor does the capacitance increase? Explain.

24. The plates of a parallel-plate capacitor are connected to a battery. If the battery voltage is doubled, by what factor does the energy stored in the capacitor increase?

25. A parallel-plate capacitor is charged, and then disconnected from the battery that charged it; both plates are now electrically isolated. The capacitor charge, capacitance, and potential difference are Q_i, C_i, and $(\Delta V_C)_i$. Then a dielectric is inserted between the plates. Afterward, the charge, capacitance, and potential difference are Q_f, C_f, and $(\Delta V_C)_f$.

a. Is Q_f larger than, smaller than, or the same as Q_i? Explain.

b. Is C_f larger than, smaller than, or the same as C_i? Explain.

c. Is $(\Delta V_C)_f$ larger than, smaller than, or the same as $(\Delta V_C)_i$? Explain.

26. Rank in order, from largest to smallest, the energies $(U_C)_1$ to $(U_C)_4$ stored in each of these capacitors.

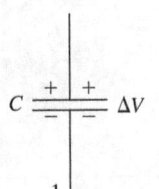

$C \quad \Delta V$ 1

$\frac{1}{2}C \quad 2\Delta V$ 2

$2C \quad \frac{1}{2}\Delta V$ 3

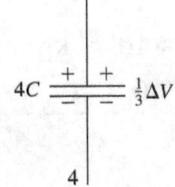

$4C \quad \frac{1}{3}\Delta V$ 4

Order:

Explanation:

You Write the Problem!

Exercises 27–30: You are given the equation that is used to solve a problem. For each of these:

    a. Write a *realistic* physics problem for which this is the correct equation. Look at worked examples and end-of-chapter problems in the textbook to see what realistic physics problems are like.

    b. Finish the solution of the problem.

27. $\dfrac{(9.0 \times 10^9 \text{ N} \cdot \text{m}^2/\text{C}^2) \, q_1 q_2}{0.030 \text{ m}} = 9.0 \times 10^{-5} \text{ J}$

    $q_1 + q_2 = 40 \text{ nC}$

28. $\dfrac{(9.0 \times 10^9 \text{ N} \cdot \text{m}^2/\text{C}^2)(3.0 \times 10^{-9} \text{ C})}{r} = 18{,}000 \text{ V}$

29. $\frac{1}{2}(1.67 \times 10^{-27} \text{ kg}) \, v_i^2 + 0 =$

    $0 + \dfrac{(9.0 \times 10^9 \text{ N} \cdot \text{m}^2/\text{C}^2)(2.0 \times 10^{-9} \text{ C})(1.60 \times 10^{-19} \text{ C})}{0.0010 \text{ m}}$

30. $400 \text{ nC} = (100 \text{ V}) \, C$

    $C = \dfrac{(8.85 \times 10^{-12} \text{ F/m})(0.10 \text{ m} \times 0.10 \text{ m})}{d}$

22 Current and Resistance

22.1 A Model of Current

22.2 Defining and Describing Current

1. a. Describe an experiment that provides evidence that current consists of charge flowing through a conductor. Use both pictures and words.

   b. One model of current is that it consists of the motion of discrete charged particles. Another model is that current is the flow of a continuous charged fluid. Does the experiment you described in part a provide evidence in favor of either one of these models? If so, describe how.

2. Figure A shows capacitor plates that have been charged to $\pm Q$. A very long conducting wire is then connected to the plates as shown in Figure B. Will the separated charges remain on the plates where they are attracted to each other by Coulomb attraction, or will the charges travel through the long wire to discharge the plates? Explain.

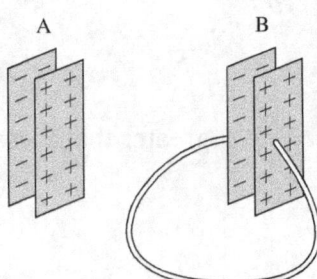

3. A discharging capacitor is used to light bulbs of different types and at different locations as shown.

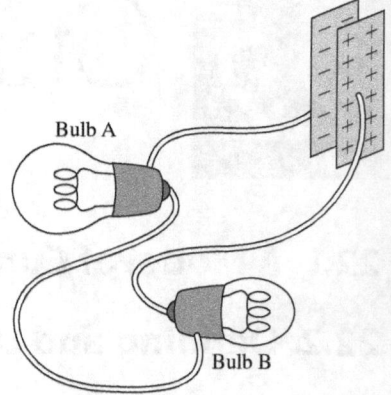

a. Which of the following, if any, will cause the current in Bulb A to be different than in Bulb B? Explain.
   - Bulb A is rated at a higher wattage than Bulb B.
   - Bulb A is connected directly to the negative plate, while Bulb B is connected to the positive plate.
   - Bulb A is closer (connected through less wire) to the capacitor plates than Bulb B.
   - None of these.

b. Which of the following, if any, will cause Bulb A to light before Bulb B? Which will cause Bulb B to light before Bulb A? Explain.
   - Bulb A is rated at a higher wattage than Bulb B.
   - Bulb A is connected directly to the negative plate, while Bulb B is connected to the positive plate.
   - Bulb A is closer to the capacitor plates than Bulb B.
   - None of these.

4. Is I_2 greater than, less than, or equal to I_1? Explain.

5. All wires in this figure are made of the same material and have the same diameter. Rank in order, from largest to smallest, the currents I_1 to I_4.

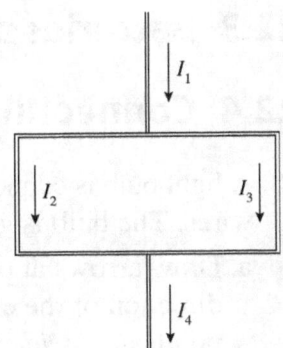

> Order:
>
> Explanation:

6. A wire carries a 4 A current. What is the current in the second wire that delivers twice as much charge in half the time?

7. What is the size of the current in the fourth wire? Is the current into or out of the junction? Explain.

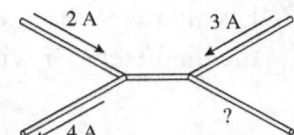

22.3 Batteries and emf

22.4 Connecting Potential and Current

8. A light bulb is connected to a battery with 1-mm-diameter wires. The bulb is glowing.

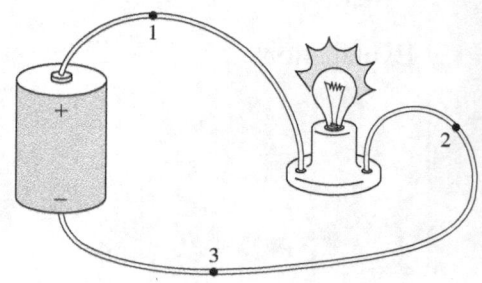

a. Draw arrows at points 1, 2, and 3 to show the direction of the electric field at those points. (The points are *inside* the wire.)

b. Rank in order, from largest to smallest, the currents I_1, I_2, and I_3 at points 1 to 3.

Order:

Explanation:

9. You have a long wire with resistance R. You would like to have a wire of the same length but resistance $2R$. Should you (a) change to a wire of the same diameter but made of a material having twice the resistivity, or (b) change to a wire made of the same material but with half the diameter? Or will either do? Explain.

10. Wire 1 and wire 2 are made of the same metal and are the same length. Wire 1 has twice the diameter and half the potential difference across its ends. What is the ratio of I_1/I_2?

22.5 Ohm's Law and Resistor Circuits

11. A wire consists of two equal-diameter segments. Their resistivities differ, with $\rho_1 > \rho_2$. The current in segment 1 is I_1. Compare the values of the currents in the two segments. Is I_2 greater than, less than, or equal to I_1? Explain.

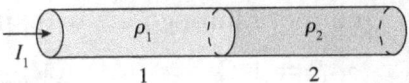

12. A graph of current as a function of potential difference is given for a particular wire segment.

    a. What is the resistance of the wire?

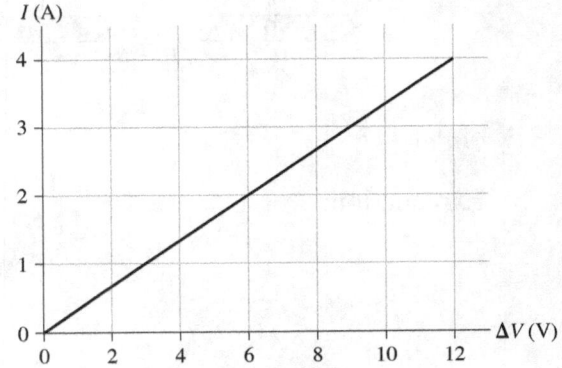

    b. Sketch and label on the same axes the graph of current vs. potential difference for a wire made of the same material but twice as long as the wire in part a.
    c. Sketch and label on the same axes the graph of current vs. potential difference for a wire made of the same material but with twice the cross-section area of the wire in part a.

13. For resistors R_1 and R_2:

    a. Which end (left, right, top, or bottom) is more positive?

    R_1: _____ R_2: _____

    b. In which direction (such as left to right or top to bottom) does the potential decrease?

    R_1: _____

    R_2: _____

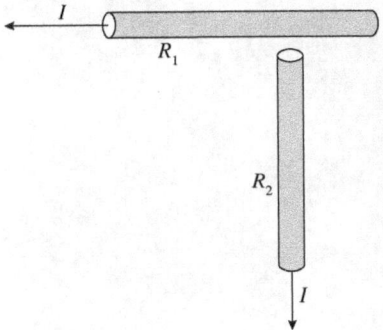

14. A wire is connected to the terminals of a 6 V battery. What is the potential difference ΔV_{wire} between the ends of the wire, and what is the current I through the wire, if the wire has the following resistances:

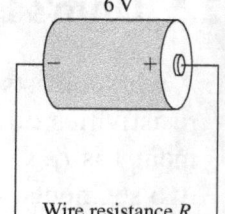

Wire resistance R

   a. $R = 1 \ \Omega$ $\Delta V_{\text{wire}} = $ _____ $I = $ _____

   b. $R = 2 \ \Omega$ $\Delta V_{\text{wire}} = $ _____ $I = $ _____

   c. $R = 3 \ \Omega$ $\Delta V_{\text{wire}} = $ _____ $I = $ _____

   d. $R = 6 \ \Omega$ $\Delta V_{\text{wire}} = $ _____ $I = $ _____

15. Rank in order, from largest to smallest, the currents I_1 to I_4 through these four resistors.

+ 2 V −	+ 1 V −	+ 2 V −	+ 1 V −
2 Ω	2 Ω	1 Ω	1 Ω
I_1	I_2	I_3	I_4

Order:

Explanation:

22.6 Energy and Power

16. a. Two conductors of equal lengths are connected in a line so that the same current I flows through both. The conductors are made of the same material but have different radii. Which of the two conductors dissipates the larger amount of power? Explain.

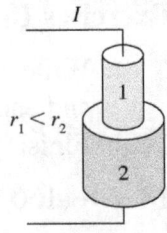

b. The same two conductors in part a are then each connected separately to identical batteries so that each has the same potential difference across it. Which of the two conductors now dissipates the larger amount of power? Explain.

17. Two resistors of equal lengths are connected to a battery by ideal wires. The resistors have the same radii but are made of different materials and have different resistivities ρ with $\rho_1 > \rho_2$.

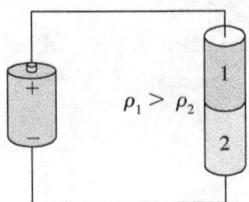

a. Is the current I_1 in resistor 1 larger than, smaller than, or the same as I_2 in resistor 2? Explain.

b. Which of the two resistors dissipates the larger amount of power? Explain.

c. Is the voltage ΔV_1 across resistor 1 larger than, smaller than, or the same as ΔV_2 across resistor 2? Explain.

You Write the Problem!

Exercises 18–20: You are given the equation that is used to solve a problem. For each of these:

    a. Write a *realistic* physics problem for which this is the correct equation. Look at worked examples and end-of-chapter problems in the textbook to see what realistic physics problems are like.

    b. Finish the solution of the problem.

18. $I \times (150\ \Omega) = 4.5\ \text{V}$

19. $100\ \text{W} = \dfrac{(120\ \text{V})^2}{R}$

20. $(2.0\ \text{A}) \times \dfrac{(1.5 \times 10^{-6}\ \Omega \cdot \text{m})L}{\pi\,(7.5 \times 10^{-4}\ \text{m})^2} = 12\ \text{V}$

23 Circuits

23.1 Circuit Elements and Diagrams

23.2 Kirchhoff's Laws

Exercises 1–4: Redraw the circuits shown using standard circuit symbols with only right angle corners.

1.

2.

3.

4.

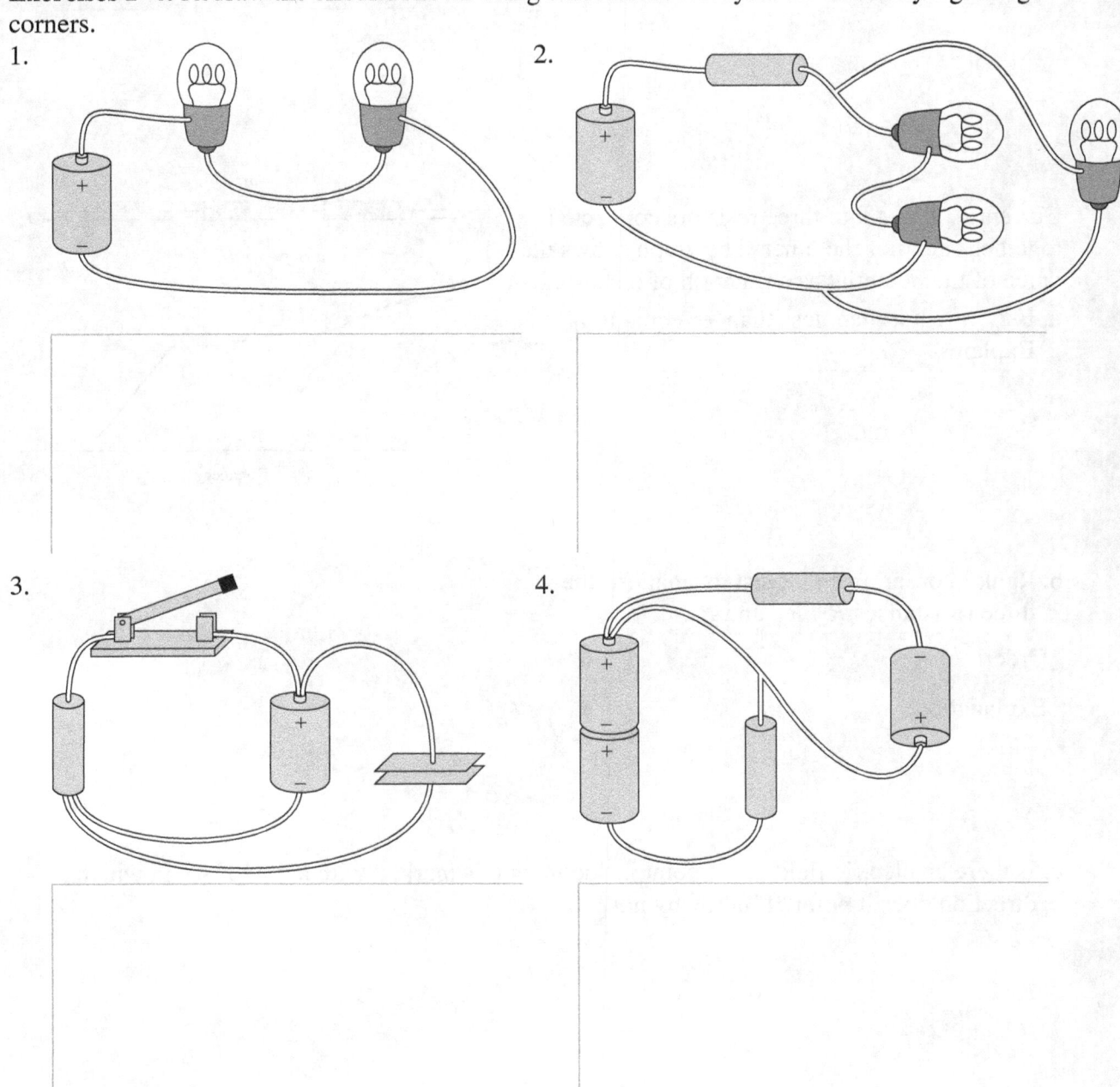

5. The tip of a flashlight bulb is touching the top of a 3 V battery. Does the bulb light? Why or why not?

6. A flashlight bulb is connected between two 1.5 V batteries as shown. Does the bulb light? Why or why not?

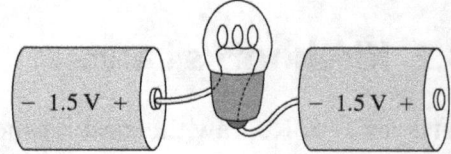

7. Current I_{in} flows into three resistors connected together one after the other. The graph shows the value of the potential as a function of distance.

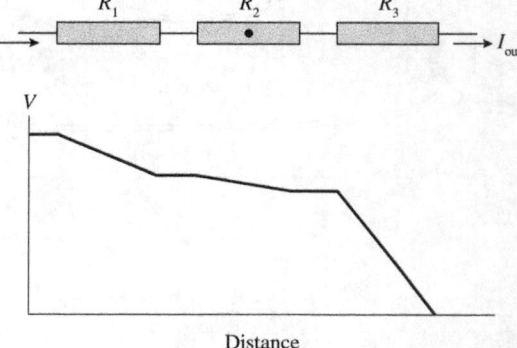

a. Is I_{out} greater than, less than, or equal to I_{in}? Explain.

b. Rank in order, from largest to smallest, the three resistances R_1, R_2, and R_3.

Order:

Explanation:

c. Is there an electric field at the point inside R_2 that is marked with a dot? If so, in which direction does it point? If not, why not?

8. a. Redraw the circuit shown as a standard circuit diagram.

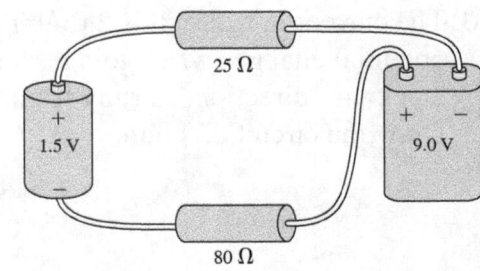

b. Assign a direction for the current and label the current arrow I on your sketch.

c. Apply Kirchhoff's loop law to determine the current through the resistors.

9. Draw a circuit for which the Kirchhoff loop law equation is

$$6V - I \cdot 2\Omega + 3V - I \cdot 4\Omega = 0$$

Assume that the analysis is done in a clockwise direction.

10. The current in a circuit is 2.0 A. The graph shows how the potential changes when going around the circuit in a clockwise direction, starting from the lower left corner. Draw the circuit diagram.

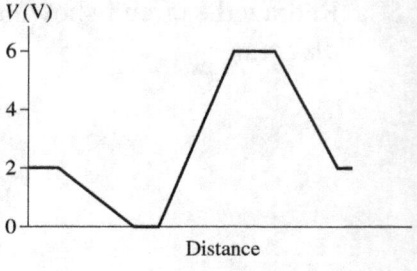

23.3 Series and Parallel Circuits

11. What is the equivalent resistance of each group of resistors?

a.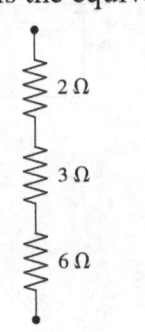

b.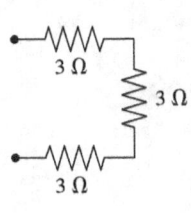

c.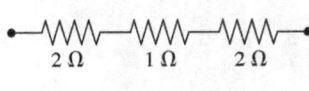

$R_{eq} =$ _____ $R_{eq} =$ _____ $R_{eq} =$ _____

12. What is the equivalent resistance of each group of resistors?

a.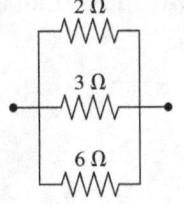

b.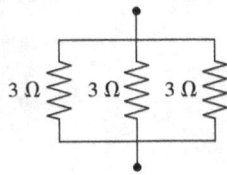

c.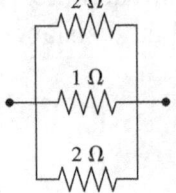

$R_{eq} =$ _____ $R_{eq} =$ _____ $R_{eq} =$ _____

13. The figure shows five combinations of identical resistors. Rank in order, from largest to smallest, the equivalent resistances $(R_{eq})_1$ to $(R_{eq})_5$.

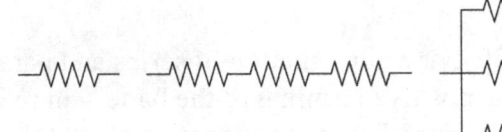   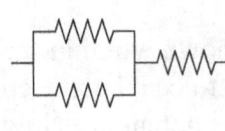

    1   2   3   4   5

Order:

Explanation:

14. A 60 W light bulb and a 100 W light bulb are placed one after the other in a circuit. The battery's emf is large enough that both bulbs are glowing. Which one glows more brightly? Explain.

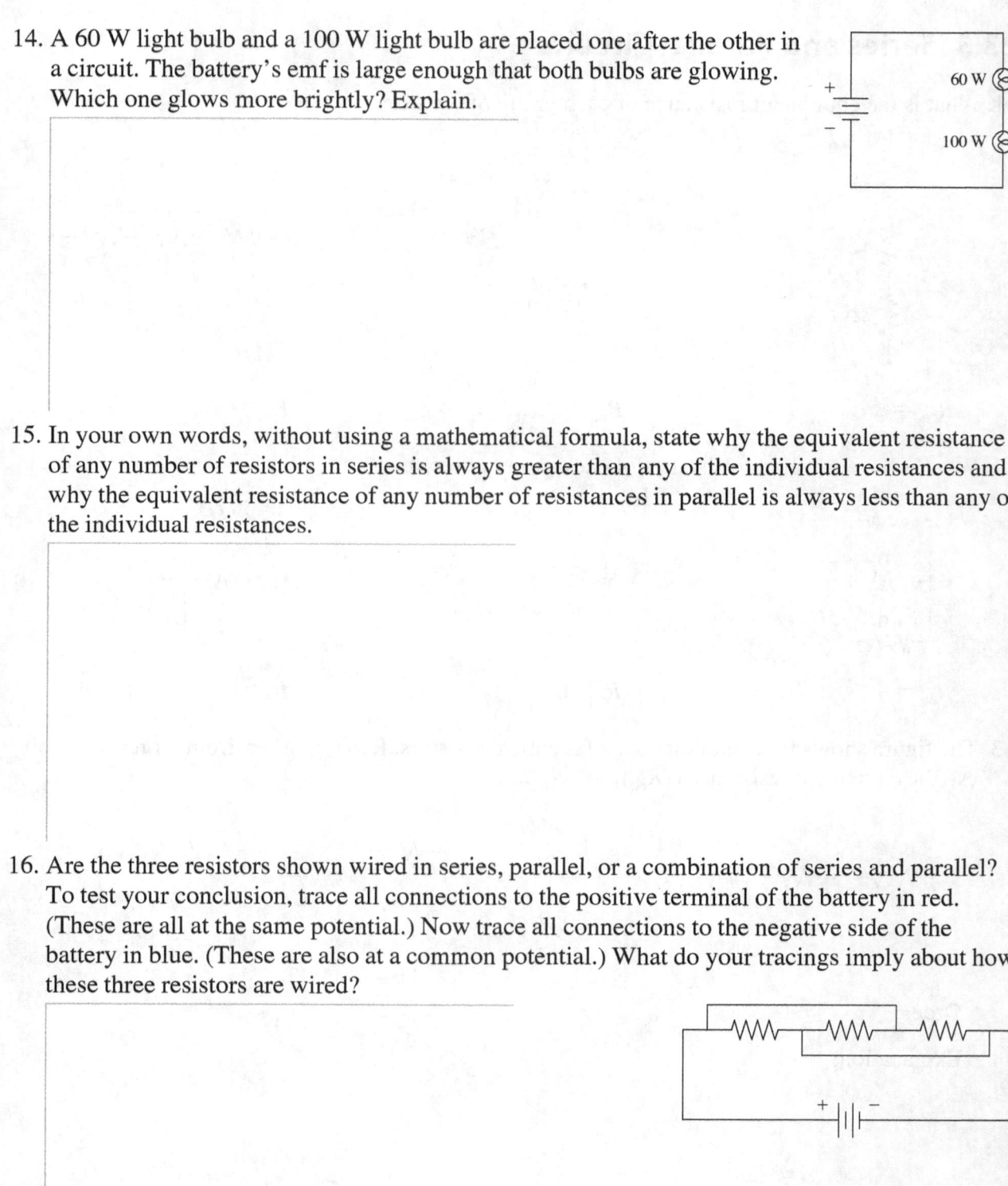

15. In your own words, without using a mathematical formula, state why the equivalent resistance of any number of resistors in series is always greater than any of the individual resistances and why the equivalent resistance of any number of resistances in parallel is always less than any of the individual resistances.

16. Are the three resistors shown wired in series, parallel, or a combination of series and parallel? To test your conclusion, trace all connections to the positive terminal of the battery in red. (These are all at the same potential.) Now trace all connections to the negative side of the battery in blue. (These are also at a common potential.) What do your tracings imply about how these three resistors are wired?

23.4 Measuring Voltage and Current

17. A student assigned the task of measuring the current through the 50 Ω resistor sets up the circuit shown.

    a. Why is this method wrong?

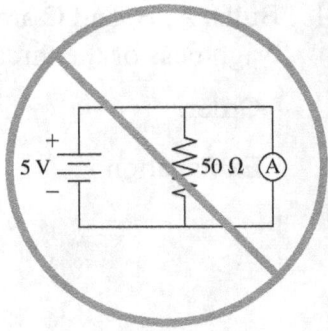

    b. Draw a circuit to show how the current *should* be measured.

18. A student assigned the task of measuring the voltage across the 2 Ω resistor sets up the circuit shown.

    a. Why is this method wrong?

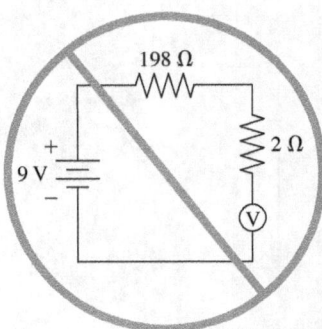

    b. Draw a circuit to show how the voltage *should* be measured.

23.5 More Complex Circuits

19. Bulbs A, B, and C are identical. Rank in order, from most to least, the brightness of the three bulbs.

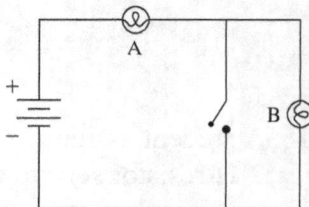

Order:

Explanation:

20. Initially bulbs A and B are glowing. Then the switch is closed. What happens to each bulb? Does it get brighter, stay the same, get dimmer, or go out? Explain your reasoning.

21. Bulbs A and B are identical. Initially, both are glowing.

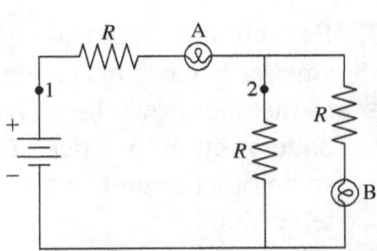

   a. Bulb A is removed from its socket. What happens to bulb B? Does it get brighter, stay the same, get dimmer, or go out? Explain.

   b. Bulb A is replaced. Bulb B is then removed from its socket. What happens to bulb A? Does it get brighter, stay the same, get dimmer, or go out? Explain.

   c. The circuit is restored to its initial condition. A wire is then connected between points 1 and 2. What happens to the brightness of each bulb?

22. Two batteries are identical and the four resistors all have exactly the same resistance.

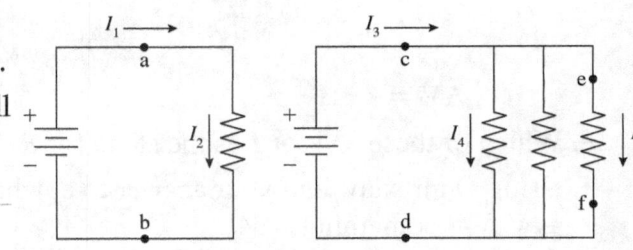

   a. Compare ΔV_{ab}, ΔV_{cd}, and ΔV_{ef}. Are they all the same? If not, rank them in decreasing order. Explain your reasoning.

   b. Rank in order, from largest to smallest, the five currents I_1 to I_5.

Order:

Explanation:

23. Real circuits frequently have a need to reduce a voltage to a
PSA smaller value. This is done with a *voltage-divider circuit* such
23.1 as the one shown here. We've shown the input voltage V_{in} as a
battery, but in practice, it might be a voltage signal produced by
some other circuit, such as the receiver circuit in your
television.

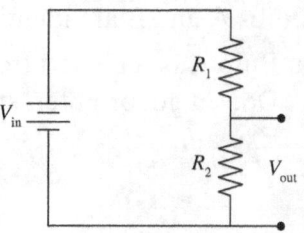

a. Is this a series or a parallel circuit? _____

b. Symbolically, what is the value of the equivalent resistance? $R_{eq} =$ _____

c. Redraw the circuit with the two resistors replaced with the equivalent resistance. Note that
V_{out} will not be seen in this circuit.

d. You now have one resistor connected directly across the battery. Find the voltage across
and the current through this resistor. Write your answers in terms of V_{in}, R_1, and R_2.

$\Delta V =$ _____ $I =$ _____

e. Which of these, ΔV or I, is the same for R_1 and R_2 as for R_{eq}? _____

f. Using Ohm's law and your answer to e, what is V_{out}? It's most useful to write your answer
as $V_{out} =$ something $\times V_{in}$.

$V_{out} =$ _____

g. What is V_{out} if $R_2 = 2R_1$? $V_{out} =$ _____

23.6 Capacitors in Parallel and Series

24. Each capacitor in the circuits below has capacitance C. What is the equivalent capacitance of the group of capacitors?

a. 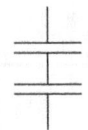 $C_{eq} = $ _____

b. 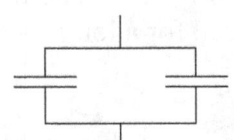 $C_{eq} = $ _____

c. 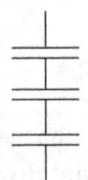 $C_{eq} = $ _____

d. 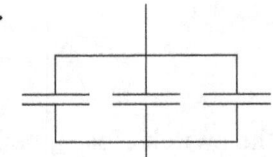 $C_{eq} = $ _____

e. 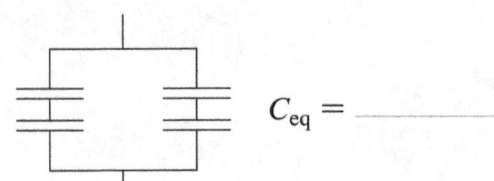 $C_{eq} = $ _____

f. 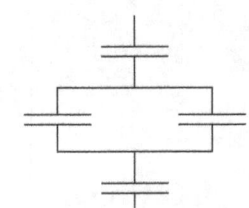 $C_{eq} = $ _____

25. Rank in order, from largest to smallest, the equivalent capacitances $(C_{eq})_1$ to $(C_{eq})_4$ of these four groups of capacitors.

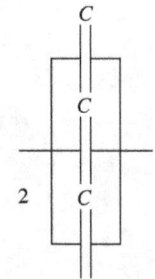

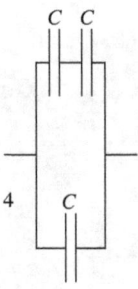

Order:

Explanation:

23.7 *RC* Circuits

26. The charge on the capacitor is zero when the switch closes at $t = 0$ s.

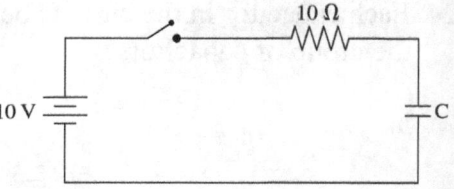

a. What will be the current in the circuit after the switch has been closed for a long time? Explain.

b. Immediately after the switch closes, before the capacitor has had time to charge, the potential difference across the capacitor is zero. What must be the potential difference across the resistor in order to satisfy Kirchhoff's loop law? Explain.

c. Based on your answer to part b, what is the current in the circuit immediately after the switch closes?

d. Sketch a graph of current versus time, starting from just before $t = 0$ s and continuing until the switch has been closed a long time. There are no numerical values for the horizontal axis, so you should think about the *shape* of the graph.

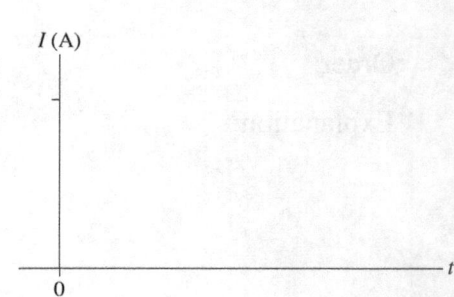

27. Graphs below show voltage versus time for the capacitor and the resistor in the *RC* circuit at the right.

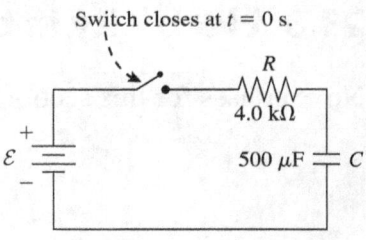

Switch closes at $t = 0$ s.

a. Write the appropriate times in the blanks below the horizontal axes of the first two graphs. Don't forget to include units.

b. Complete the third graph by showing the *sum* of the voltages across the resistor and the capacitor and by writing the appropriate times in the blanks below the axis.

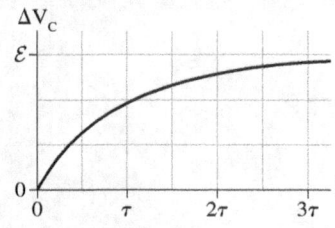
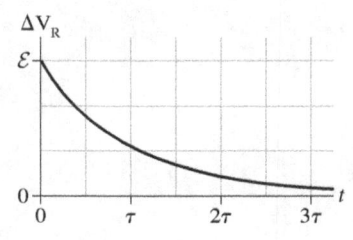
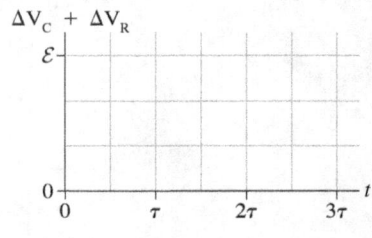

28. The capacitors in each circuit are discharged when the switch closes at $t = 0$ s. Rank in order, from largest to smallest, the time constants τ_1 to τ_5 with which each circuit will discharge.

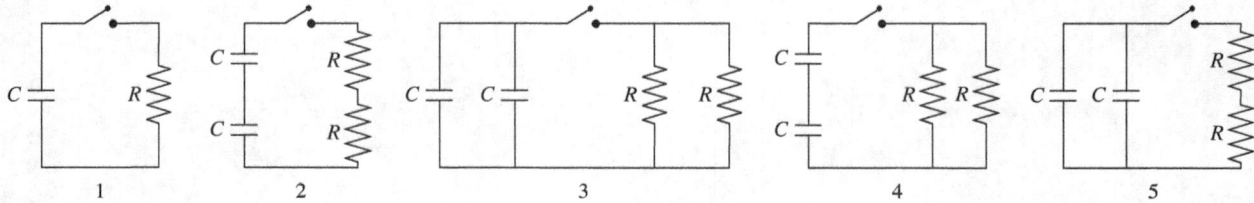

Order:

Explanation:

23.8 Electricity in the Nervous System

No exercises for this section.

24 Magnetic Fields and Forces

24.1 Magnetism

1. The compass needle below is free to rotate in the plane of the page. Either a bar magnet or a charged rod is brought toward the *center* of the compass. Does the compass rotate? If so, does it rotate clockwise or counterclockwise? If not, why not?

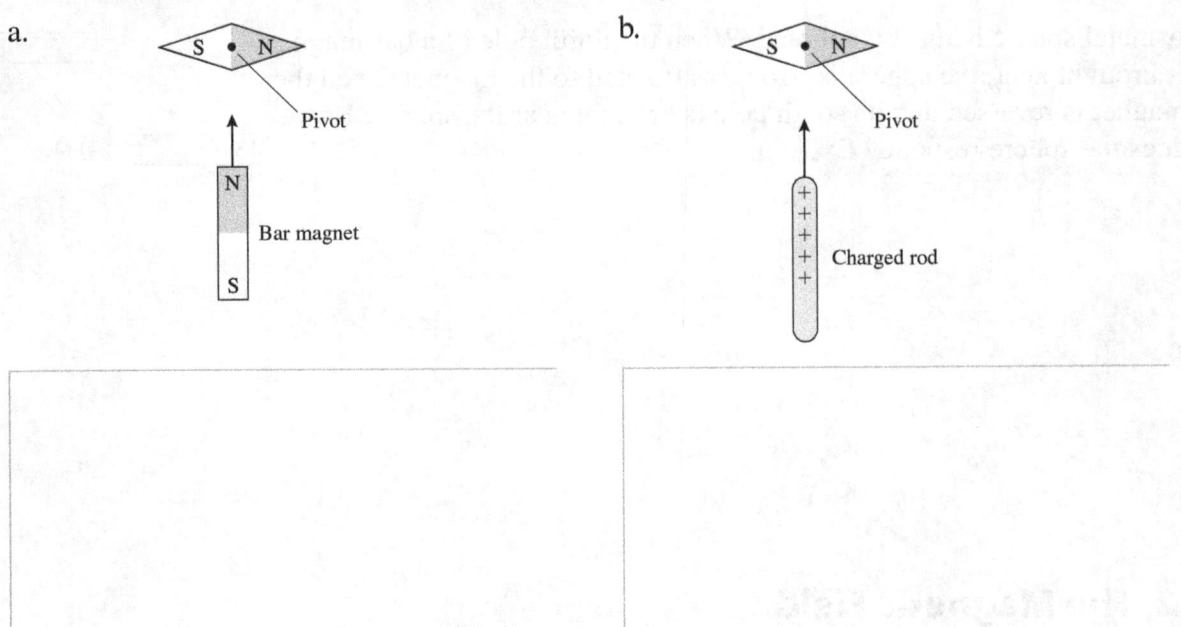

a.

b.

2. You have two electrically neutral metal cylinders that exert strong attractive forces on each other. You have no other metal objects. Can you determine if *both* of the cylinders are magnets, or if one is a magnet and the other just a piece of iron? If so, how? If not, why not?

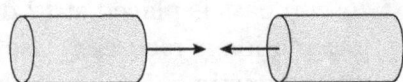

3. Can you think of any kind of object that is repelled by *both* ends of a bar magnet? If so, what? If not, what prevents this from happening?

4. A metal sphere hangs by a thread. When the north pole of a bar magnet is brought near, the sphere is strongly attracted to the magnet. Then the magnet is reversed and its south pole is brought near the sphere. How does the sphere respond? Explain.

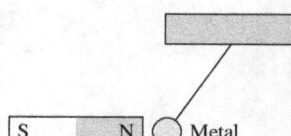

24.2 The Magnetic Field

5. A compass is placed at 12 different positions and its orientation is recorded. Use this information to draw the magnetic *field lines* in this region of space. Draw the field lines on the figure.

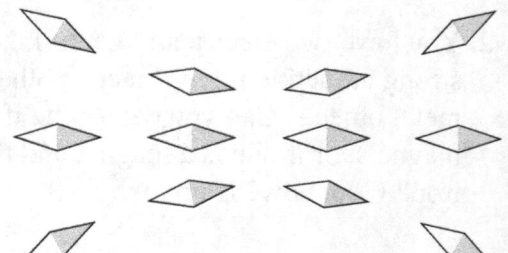

24.3 Electric Currents Also Create Magnetic Fields

6. A neutral copper rod, a polarized insulator rod, and a bar magnet are arranged around a current-carrying wire as shown. For each, will it stay where it is? Move toward or away from the wire? Rotate clockwise or counterclockwise? Explain.

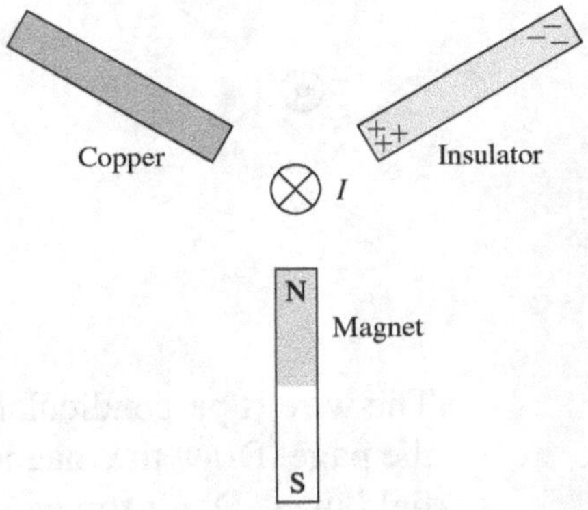

   a. | Neutral copper rod:

   b. | Insulating rod:

   c. | Bar magnet:

7. For each of the current-carrying wires shown, draw a compass needle in its equilibrium orientation at the positions of the dots. Label the poles of the compass needle.

   a.

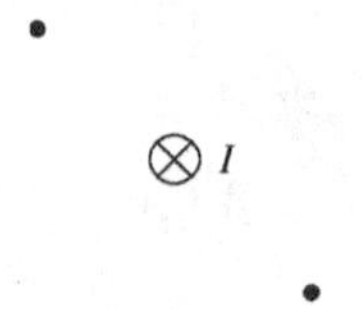

   b.

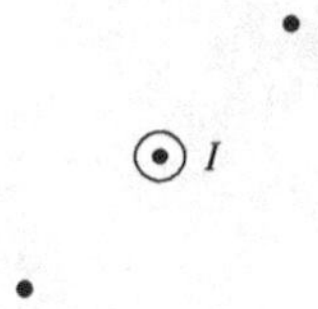

8. The figure shows a current-carrying wire directed into the page and a nearby compass needle. Is the wire's current going into the page or coming out of the page? Explain.

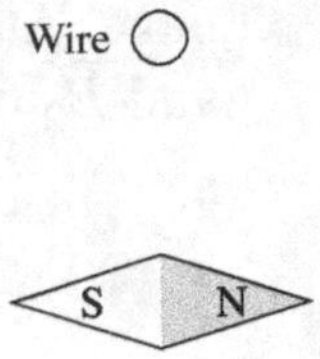

9. Each figure below shows a current-carrying wire. Draw the magnetic field.

a.   b.

The wire is perpendicular to the page. Draw the magnetic field lines. Don't forget to include arrows to show the field direction.

The wire is in the plane of the page. Use crosses and dots to show the magnetic field above and below the wire.

10. This current-carrying wire is in the plane of the page. Show the magnetic field on both sides of the wire.

11. Use an arrow to show the current direction in this wire.

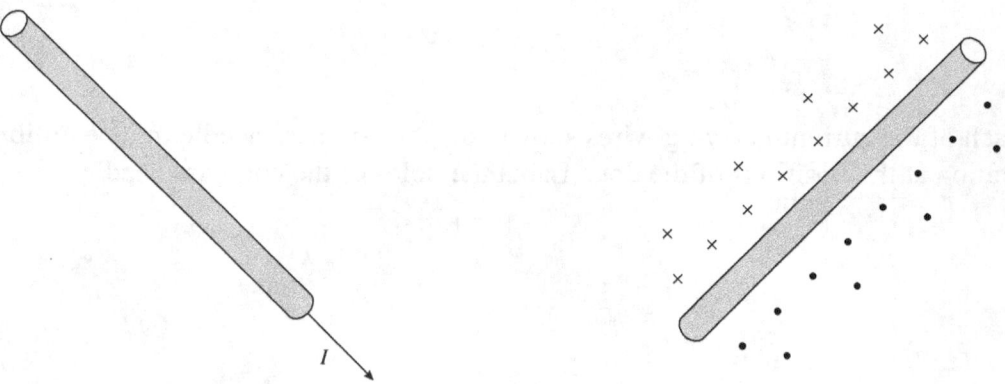

12. Each figure below shows two long straight wires carrying equal currents in and out of the page. At each of the dots, use a **black** pen or pencil to show and label the magnetic fields $\vec{B}_1$ and $\vec{B}_2$ of each wire. Then use a **red** pen or pencil to show the net magnetic field.

a.   b.

24.4 Calculating the Magnetic Field Due to a Current

13. A long straight wire, perpendicular to the page, passes through a uniform magnetic field. The *net* magnetic field at point 3 is zero.

    a. On the figure, show the direction of the current in the wire.

    b. Points 1 and 2 are the same distance from the wire as point 3; point 4 is twice as distant. Construct vector diagrams at points 1, 2, and 4 to determine the net magnetic field at each point.

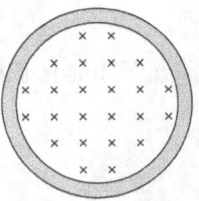

14. The figure shows the magnetic field seen when facing a current loop in the plane of the page. On the figure, show the direction of the current in the loop.

15. Rank in order, from largest to smallest, the magnetic field strengths B_1 to B_3 produced by these three solenoids.

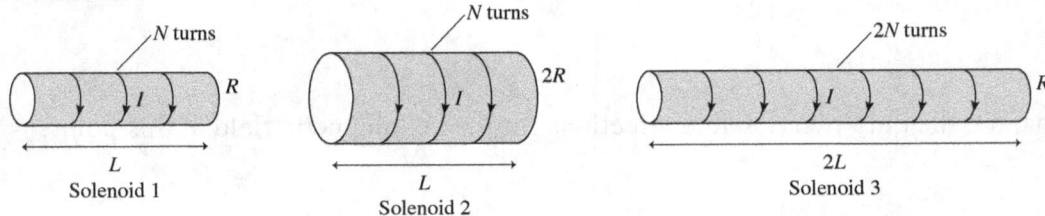

Order:

Explanation:

16. Suppose you need to find the magnetic field near the intersection of two long, straight, current-carrying wires. Assume that one wire lies directly on top of the other. Let the intersection of the wires be the origin of a coordinate system and let the point of interest, which is in the same plane, have coordinates (x, y). Recall that the magnetic field is a vector, having both a magnitude and a direction.

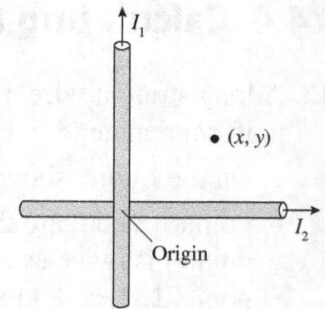

a. At the point of interest, what is the direction of magnetic field $\vec{B}_1$ due to current I_1? Explain.

b. Write an expression for the magnitude of $\vec{B}_1$.

c. What is the direction of magnetic field $\vec{B}_2$ due to current I_2? Explain.

d. Write an expression for the magnitude of $\vec{B}_2$.

e. What are the only two possible directions for the net magnetic field at this point?

f. Would knowing $I_1 > I_2$ be enough information to determine the direction of the net magnetic field? Why or why not?

g. Let a magnetic field pointing out of the page have a positive value, one pointing into the page a negative value. This is actually B_z, the magnetic field along the z-axis. Use your results for parts a–d to write an expression for B_z at position (x, y). This will be a symbolic expression in terms of quantities defined on the figure.

24.5 Magnetic Fields Exert Forces on Moving Charges

17. For each of the following, draw the magnetic force vector on the charge or, if appropriate, write "$\vec{F}$ into page," "$\vec{F}$ out of page," or "$\vec{F} = \vec{0}$."

a. 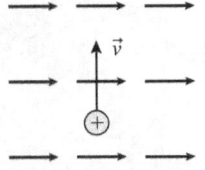   b. 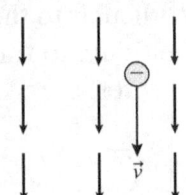   c.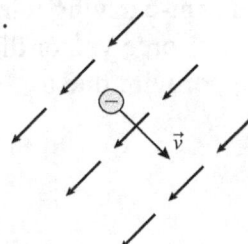

d.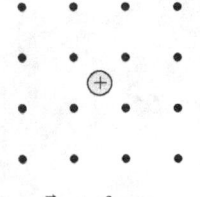

$\vec{v}$ out of page

e. 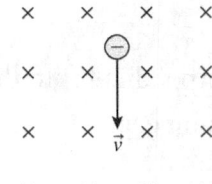   f.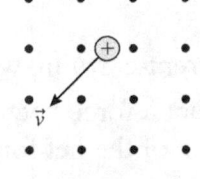

18. For each of the following, is the charge positive or negative? Write a + or a − on the charge.

a.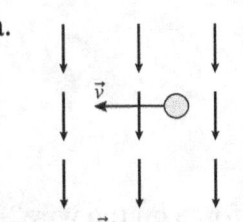

$\vec{F}$ into page

b.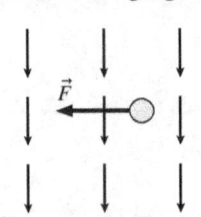

$\vec{v}$ into page

c. 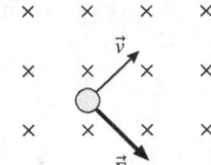   d.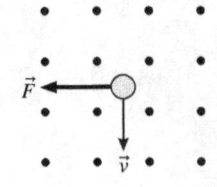

19. A positive ion, initially traveling into the page, is shot through the gap in a magnet. Is the ion deflected up, down, left, or right? Explain.

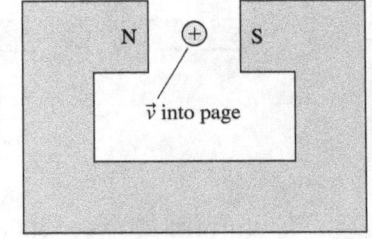

$\vec{v}$ into page

20. A positive ion is shot between the plates of a parallel-plate capacitor.

a. In what direction is the electric force on the ion?

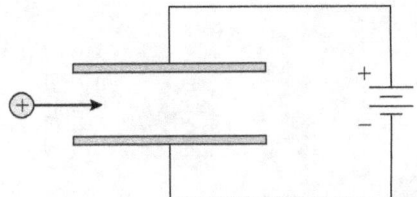

b. Could a magnetic field exert a magnetic force on the ion that is opposite in direction to the electric force? If so, show the magnetic field on the figure. If not, why not?

24.6 Magnetic Fields Exert Forces on Currents

24.7 Magnetic Fields Exert Torques on Dipoles

21. Three current-carrying wires are perpendicular to the page. Construct a force vector diagram on the figure to find the net force on the upper wire due to the two lower wires.

22. Three current-carrying wires are perpendicular to the page.

    a. Construct a force vector diagram on each wire to determine the direction of the net force on each wire.

    b. Can three *charges* be placed in a triangular pattern so that their force diagram looks like this? If so, draw it. If not, why not?

23. A current-carrying wire passes between two bar magnets. Is there a net force on the wire? If so, draw the force vector. If not, why not?

    a.

    | S | N | ⊗ | S | N |

    b.

    | S | N | ⊗ | N | S |

24. The current loop exerts a repulsive force on the bar magnet. On the figure, show the direction of the current in the loop. Explain.

    ← N | S

25. The south pole of a bar magnet is held near a current loop. Does the bar magnet attract the loop, repel the loop, or have no effect on the loop? Explain.

26. A current loop is placed between two bar magnets. Does the loop move to the right, move to the left, rotate clockwise, rotate counterclockwise, some combination of these, or none of these? Explain.

27. A square current loop is placed in a magnetic field as shown.

    a. Does the loop undergo a displacement? If so, is it up, down, left, or right? If not, why not?

    b. Does the loop rotate? If so, which edge rotates out of the page and which edge into the page? If not, why not?

28. A long vertical wire, attached by two springs to a vertical wall,
PSA passes through a region of uniform magnetic field of height L.
24.1 We would like to know what current in the wire will cause the
springs to be stretched an amount x. Assume that something
unseen supports the weight of the wire.

a. If the springs are stretched, what is the direction of the
magnetic force on the wire? Draw it on the figure and
label it $\vec{F}_{wire}$.

b. What direction is the current I in the wire? Draw and label
it on the figure, and explain your choice.

c. Draw and label the forces of the springs on the wire. Call them each $\vec{F}_{sp}$. Make sure they
have the proper lengths relative to the force you drew in part a.

d. If the wire is in equilibrium, what is the relationship between the forces you've drawn? This
should be a mathematical statement.

e. Write expressions for any forces to the left. These will be symbolic expressions in terms of
quantities defined on the figure and in the discussion above.

f. Now write symbolic expressions for any forces to the right.

g. Insert your expressions of parts e and f into the mathematical relationship of part d.

h. Complete your solution by solving this equation for I.

You Write the Problem!

Exercises 29–32: You are given the equation that is used to solve a problem. For each of these:

    a. Write a *realistic* physics problem for which this is the correct equation. Look at worked examples and end-of-chapter problems in the textbook to see what realistic physics problems are like.

    b. Finish the solution of the problem.

29. $1.5 \times 10^{-4} \text{ T} = \dfrac{(1.257 \times 10^{-6} \text{ T} \cdot \text{m/A})(15 \text{ A})}{2\pi r}$

30. $0.015 \text{ m} = \dfrac{(9.11 \times 10^{-31} \text{ kg})(1.5 \times 10^{7} \text{ m/s})}{(1.60 \times 10^{-19} \text{ C})B}$

31. $(1.67 \times 10^{-27} \text{ kg})\, a = (1.60 \times 10^{-19} \text{ C})(7.4 \times 10^{5} \text{ m/s}) \times \dfrac{(1.257 \times 10^{-6} \text{ T} \cdot \text{m/A})(150)(2.5 \text{ A})}{0.15 \text{ m}}$

32. $m(9.8 \text{ m/s}^2) = (5.5 \text{ A})(0.075 \text{ m})(0.55 \text{ T})$

25 EM Induction and EM Waves

25.1 Induced Currents

25.2 Motional emf

1. The figures below show one or more metal wires sliding on fixed metal rails in a magnetic field. For each, determine if the induced current flows clockwise, flows counterclockwise, or is zero. Show your answer by drawing it.

a.

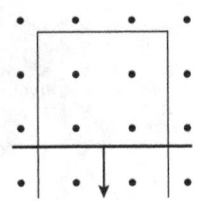

b.

c.

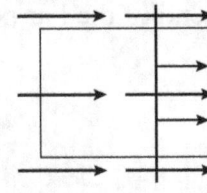

d.

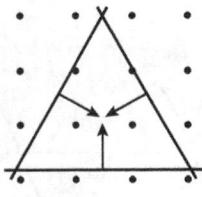

e.

f.

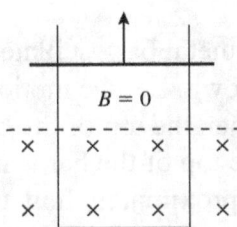

2. A loop of copper wire is being pulled from between two magnetic poles.

   a. Show on the figure the current induced in the loop. Explain your reasoning.

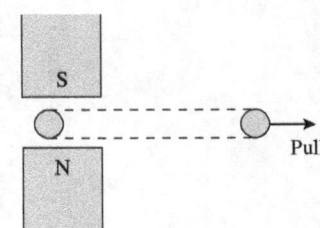

   b. Does either side of the loop experience a magnetic force? If so, draw a vector arrow or arrows on the figure to show any forces.

3. A vertical, rectangular loop of copper wire is half in and half out of a horizontal magnetic field (shaded gray). The field is zero beneath the dashed line. The loop is released and starts to fall.

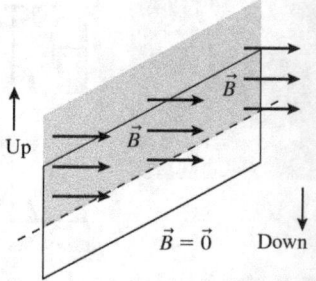

   a. Add arrows to the figure to show the direction of the induced current in the loop.

   b. Is there a net magnetic force on the loop? If so, in which direction? Explain.

4. A metal bar rotates counterclockwise in a magnetic field as shown. Does the bar have a motional emf? If not, why not? If so, is the outer end of the bar positive or negative? Explain.

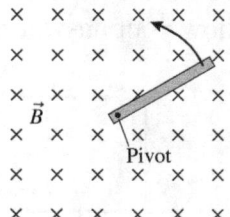

5. A metal bar oscillates back and forth on a spring as shown. Let the motional emf of the bar be positive when the top of the bar is positive and negative when the top of the bar is negative. Draw a graph showing approximately how the emf of the bar changes with time. Be sure your emf graph aligns correctly with the position graph above it.

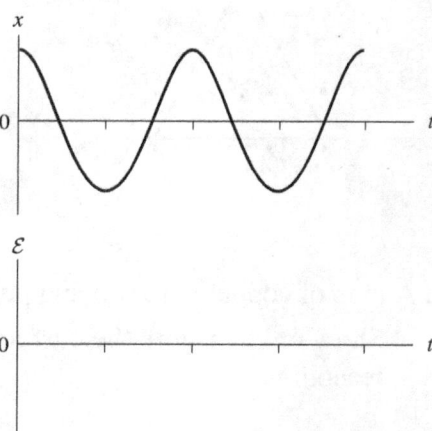

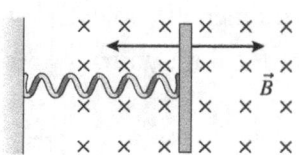

25.3 Magnetic Flux and Lenz's Law

25.4 Faraday's Law

6. The figure shows five loops in a magnetic field. The numbers indicate the lengths of the sides and the strength of the field. Rank in order, from largest to smallest, the magnetic fluxes Φ_1 to Φ_5. Some may be equal.

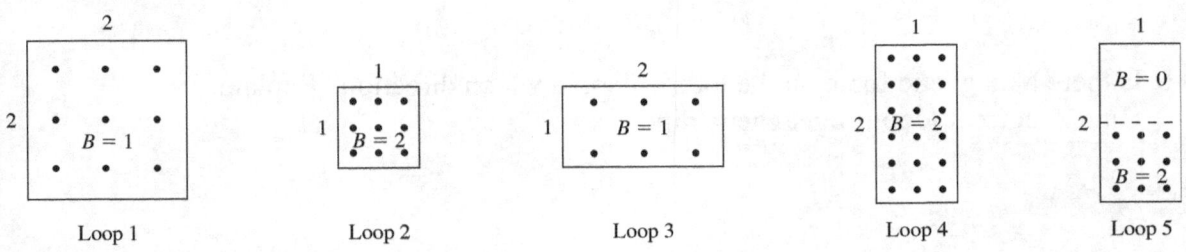

Loop 1 Loop 2 Loop 3 Loop 4 Loop 5

Order:

Explanation:

7. A circular loop rotates at constant speed about an axle through the center of the loop. The figure shows an edge view and defines the angle ϕ, which increases from $0°$ to $360°$ as the loop rotates.

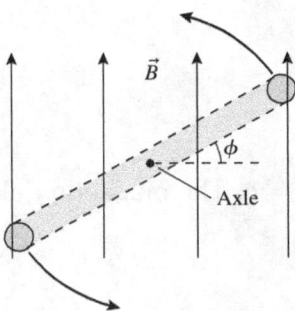

a. At what angle or angles is the magnetic flux a maximum?

b. At what angle or angles is the magnetic flux a minimum?

c. At what angle or angles is the magnetic flux *changing* most rapidly? Explain your choice.

8. A loop of wire is horizontal. A bar magnet is pushed toward the loop from below, along the axis of the loop.

   a. What is the current direction in the loop as the magnet is approaching? Explain.

   b. Is there a magnetic force on the loop? If so, in which direction? Explain.
      Hint: A current loop is a magnetic dipole.

9. Does the loop of wire have a clockwise current, a counterclockwise current, or no current under the following circumstances? Explain.

   a. The magnetic field points out of the page and its strength is increasing.

   b. The magnetic field points out of the page and its strength is constant.

   c. The magnetic field points out of the page and its strength is decreasing.

10. A loop of wire is perpendicular to a magnetic field. The magnetic field strength as a function of time is given by the top graph. Draw a graph of the current in the loop as a function of time. Let a positive current represent a current that comes out of the top of the loop and enters the bottom of the loop. There are no numbers for the vertical axis, but your graph should have the correct shape and proportions.

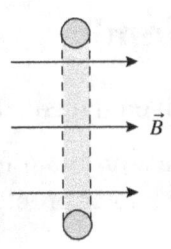
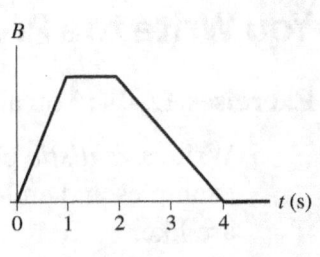
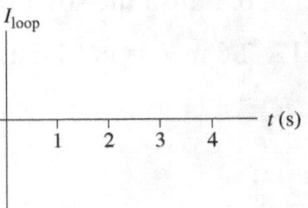

11. For the following questions, consider the current in the loop to be positive if it comes out of the top of the loop and enters the bottom; negative if it comes out of the bottom of the loop and enters the top.

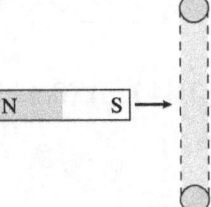

   a. As the magnet is pushed into the loop, is the current in the loop positive, negative, or zero? Explain.

   b. As the magnet is held at rest in the center of the loop, is the current in the loop positive, negative, or zero? Explain.

   c. As the magnet is withdrawn from the loop, pulled back to the left, is the current in the loop positive, negative, or zero? Explain.

   d. If the magnet is pushed into the loop more rapidly than in part a, does the size of the current increase, decrease, or stay the same? Explain.

You Write the Problem!

Exercises 12–14: You are given the equation that is used to solve a problem. For each of these:

a. Write a *realistic* physics problem for which this is the correct equation. Look at worked examples and end-of-chapter problems in the textbook to see what realistic physics problems are like.

b. Finish the solution of the problem.

12. $2.6 \text{ mV} = v(15 \text{ cm})(5.0 \times 10^{-5} \text{ T})$

13. $\pi (0.040 \text{ m})^2 (0.12 \text{ T}) \cos \theta = 5.2 \times 10^{-4} \text{ Wb}$

14. $0.25 \text{ A} = \dfrac{(0.15 \text{ m})^2 (0.30 \text{ T})/(0.50 \text{ s})}{R}$

15. The graph shows how the magnetic field changes
PSA through a rectangular loop of wire with resistance
25.1 R. Draw a graph of the current in the loop as a
function of time. Let a counterclockwise
current be positive, a clockwise current be
negative.

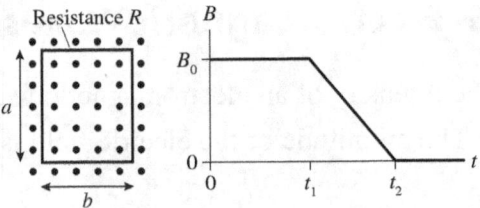

a. What is the magnetic flux through the loop at $t = 0$?

b. Does this flux *change* between $t = 0$ and $t = t_1$?

c. Is there an induced current in the loop between $t = 0$ and $t = t_1$?

d. What is the magnetic flux through the loop at $t = t_2$?

e. What is the *change* in flux through the loop between t_1 and t_2?

f. What is the time interval between t_1 and t_2?

g. What is the magnitude of the induced emf between t_1 and t_2?

h. What is the magnitude of the induced current between t_1 and t_2?

i. Does the magnetic field point out of or into the loop?

f. Between t_1 and t_2, is the magnetic flux increasing or decreasing?

g. To oppose the *change* in the flux between t_1 and t_2, should the
magnetic field of the induced current point out of or into the loop?

h. Is the induced current between t_1 and t_2 positive or negative?

i. Does the flux through the loop change after t_2?

j. Is there an induced current in the loop after t_2?

k. Use all this information to draw a graph of the induced current. Add appropriate labels on
the vertical axis.

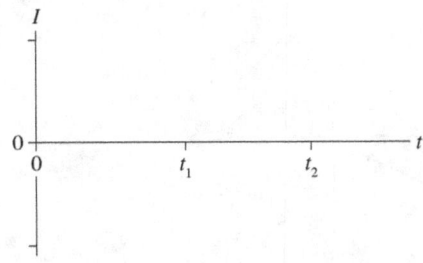

25.5 Electromagnetic Waves

16. The intensity of an electromagnetic wave is 10 W/m^2. What will be the intensity if:

    a. The amplitude of the electric field is doubled?

    b. The amplitude of the magnetic field is doubled?

    c. The frequency is doubled?

17. The intensity of a polarized electromagnetic wave is 10 W/m^2. What will be the intensity of the wave after it passes through a polarizing filter whose axis makes the following angle with the plane of polarization?

    $\theta = 0°$ _____ $\theta = 60°$ _____

    $\theta = 30°$ _____ $\theta = 90°$ _____

    $\theta = 45°$ _____

18. A polarized electromagnetic wave passes through a polarizing filter. Draw the electric field of the wave after it has passed through the filter.

    a.   b.

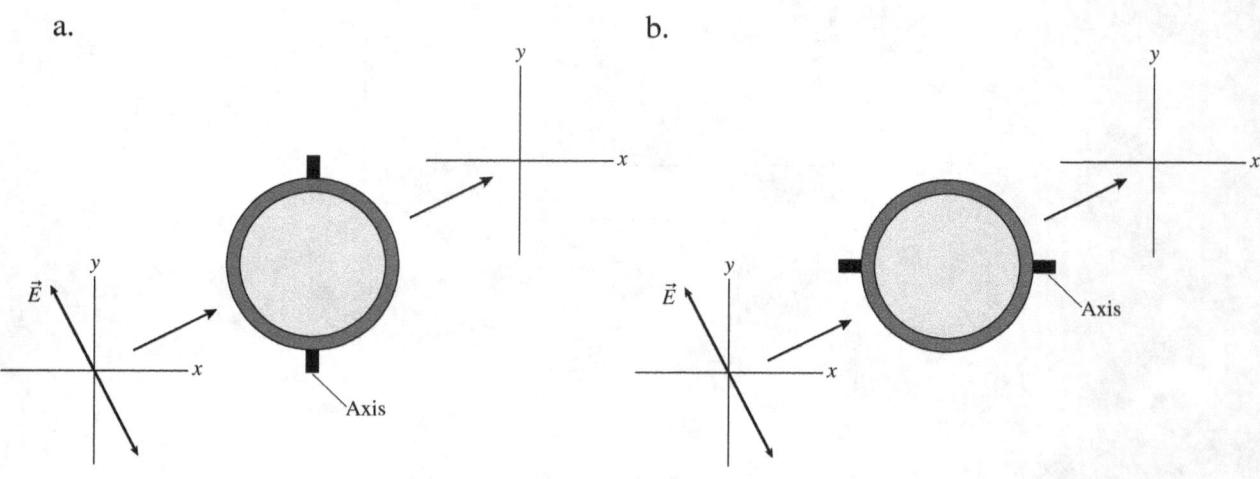

19. A polarized electromagnetic wave passes through a series of polarizing filters. Draw the electric field of the wave after it has passed through each filter or, if appropriate, state that $E = 0$.

a.

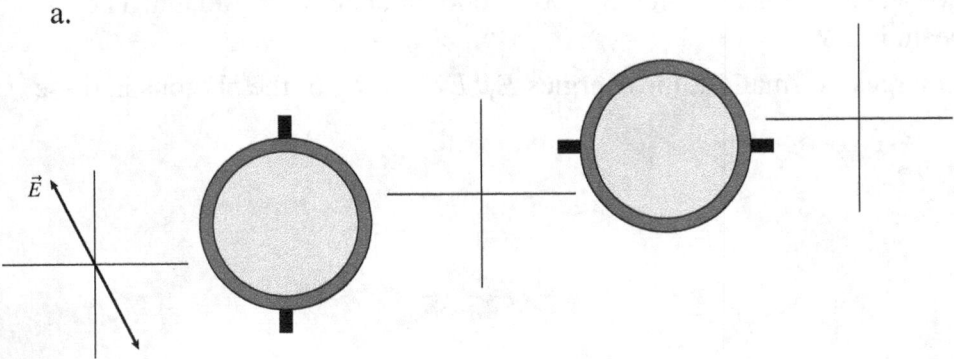

b.

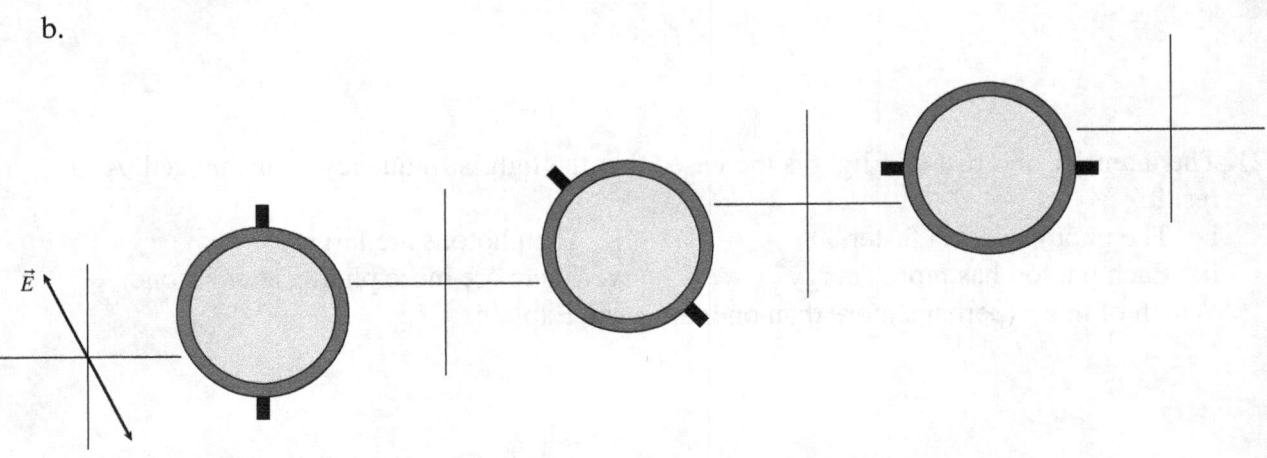

25.6 The Photon Model of Electromagnetic Waves

20. Three laser beams have wavelengths $\lambda_1 = 400$ nm, $\lambda_2 = 600$ nm, and $\lambda_3 = 800$ nm. The power of each laser beam is 1 W.

a. Rank in order, from largest to smallest, the energies E_1, E_2, and E_3 of the photons in these three laser beams.

Order:

Explanation:

b. Rank in order, from largest to smallest, the number of photons per second N_1, N_2, and N_3 delivered by the three laser beams.

Order:

Explanation:

21. The intensity of a beam of light is increased, but the light's frequency is unchanged. As a result:

i.   The photons travel faster.   iii. The photons are larger.
ii. Each photon has more energy.   iv. There are more photons per second.

Which of these (perhaps more than one) are true? Explain.

22. The frequency of a beam of light is increased, but the light's intensity is unchanged. As a result:

i.   The photons travel faster.   iii. There are fewer photons per second.
ii. Each photon has more energy.   iv. There are more photons per second.

Which of these (perhaps more than one) are true? Explain.

23. Light of wavelength $\lambda = 1$ μm is emitted from point A. A photon is detected 5 μm away at point B. On the figure, draw the trajectory that a photon follows between points A and B.

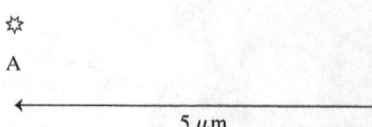

25.7 The Electromagnetic Spectrum

24. The graph at the right shows the thermal emission spectrum of light as a function of wavelength from an object at temperature T. On the graph, sketch approximately how the spectrum would appear if the object's absolute temperature were doubled to $2T$.

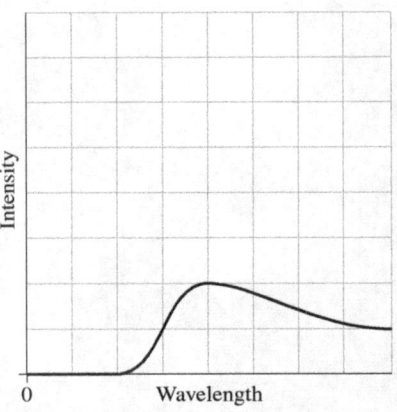

25. An object's thermal emission spectrum has peak intensity at a wavelength of 3.0 μm. What will be the wavelength of peak intensity if the object's absolute temperature is tripled?

26. An astronomer observes the thermal emission spectrum of two stars. Star A has maximum intensity at a wavelength of 400 nm. Star B has maximum intensity at 1200 nm. What is the ratio T_A/T_B of the absolute temperatures of the two stars?

26 AC Electricity

26.1 Alternating Current

1. The graph shows two cycles of the AC current in a simple AC resistor circuit.

   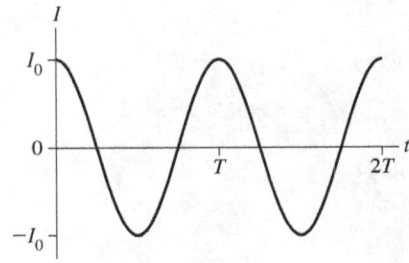

   a. What is the *average* value of the current after two whole cycles? _____

   b. On the axes below the current graph, draw a graph of the *square* of the current I^2. Make sure your graphs features are aligned with those of the AC current graph above it.

   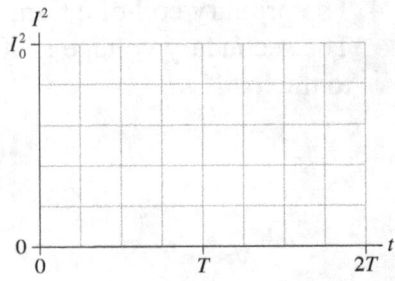

   c. If the frequency of the AC current is f, what is the frequency of the square of the current? _____

   d. Based on your graph, what is the *average* value of the square of the current after two whole cycles? _____

   e. What is the square root of your answer to part d? That is, the square root of the average of the square of the current? This value is the root-mean-square current I_{rms}. _____

2. The peak current through the resistor in an AC circuit is 4.0 A. What is the peak current if:

   a. The resistance R is doubled?

   b. The peak voltage V_R is doubled?

   c. The frequency f is doubled?

26.2 AC Electricity and Transformers

3. Your phone's 12 V battery is dead. It's a holiday and all the stores are closed. You do have two 1.5 V batteries from your camera and a transformer with $N_1 = 100$ turns on the primary and $N_2 = 400$ turns on the secondary. Can you use the batteries and transformer to get your phone operating? If so, how would you do it? If not, why not?

4. The primary coil of a transformer draws a 4.0 A rms current when plugged into a 120 V outlet. The secondary voltage is 40 V. What current does the secondary coil of the transformer deliver to the load?

26.3 Household Electricity

26.4 Biological Effects and Electrical Safety

5. The figure shows a standard North American household electric socket.

a. Label the socket holes. One is "hot," the other "neutral."
b. When you plug a device into the socket, does the current always come out of one hole and go back into the other? If so, which hole does the current come out of? If not, why not?

26.5 Capacitor Circuits

6. Current and voltage graphs are shown for a capacitor circuit with $f = 1000$ Hz.

   a. What is the capacitive reactance X_C?

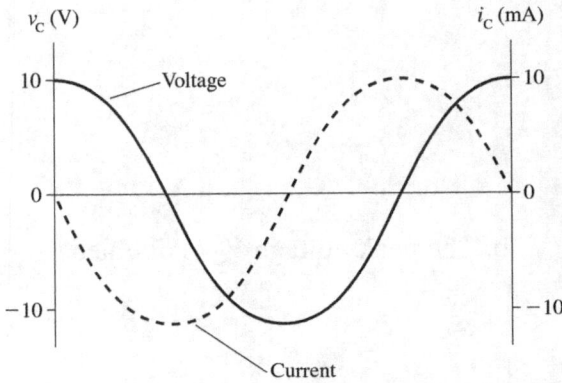

   b. What is the capacitance C?

7. Consider these three circuits.

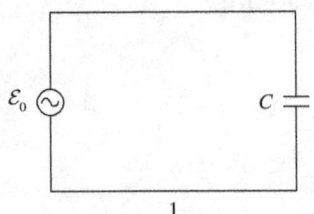 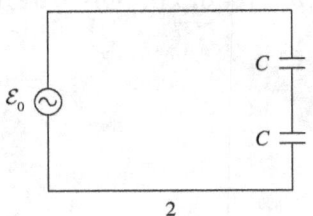 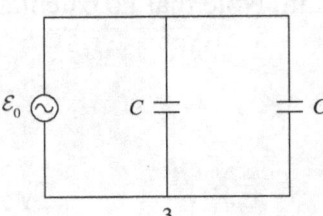

Rank in order, from largest to smallest, the peak currents $(I_C)_1$ to $(I_C)_3$.

Order:

Explanation:

8. The peak current through the capacitor in an AC circuit is 4.0 A. What is the peak current if:

   a. The capacitance C is doubled?

   b. The peak voltage V_C is doubled?

   c. The frequency f is doubled?

9. A 13 μF capacitor is connected to a 5.0 V/250 Hz oscillating voltage. What is the instantaneous capacitor current when the instantaneous capacitor voltage is $v_C = -5.0$ V? Explain. Note that no calculations are required, only careful reasoning.

26.6 Inductors and Inductor Circuits

10. Current and voltage graphs are shown for an inductor circuit with $f = 1000$ Hz.

    a. What is the inductive reactance X_L?

    b. What is the inductance L?

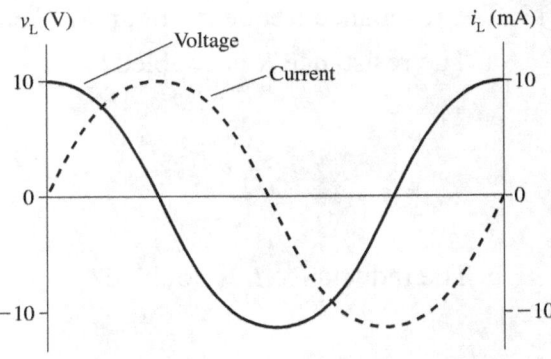

11. The peak current through the inductor in an AC circuit is 4.0 A. What is the peak current if:

    a. The inductance L is doubled?

    b. The peak voltage V_L is doubled?

    c. The frequency f is doubled?

12. Consider these three circuits.

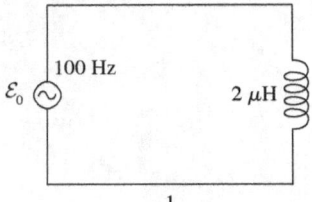 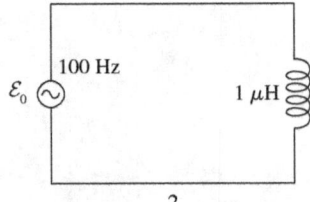 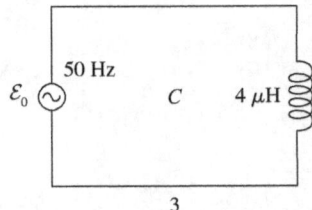

Rank in order, from largest to smallest, the peak currents $(I_L)_1$ to $(I_L)_3$.

Order:

Explanation:

26.7 Oscillation Circuits

13. The resonance frequency of an *RLC* circuit is 1000 Hz. What is the resonance frequency if:

    a. The resistance R is doubled?

    b. The inductance L is doubled?

    c. The capacitance C is doubled?

    d. The peak emf $\mathcal{E}_0$ is doubled?

14. Two *RLC* circuits have the same resistance and same capacitance but different inductances. The graph shows the current oscillation at the resonance frequency of each circuit. Is L_1 for circuit 1 larger or smaller than L_2 for circuit 2? Explain.

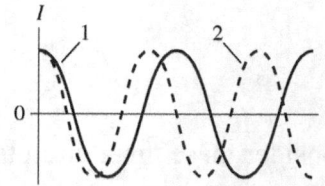

27 Relativity

27.1 Relativity: What's It All About?

27.2 Galilean Relativity

1. In which reference frame, S or S', does the ball move faster?

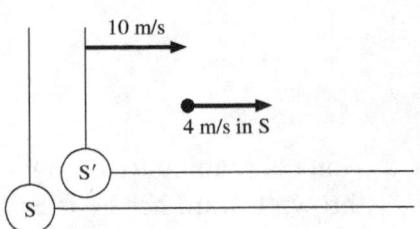

2. Frame S' moves parallel to the x-axis of frame S.

   a. Is there a value of v for which the ball is at rest in S'? If so, what is v? If not, why not?

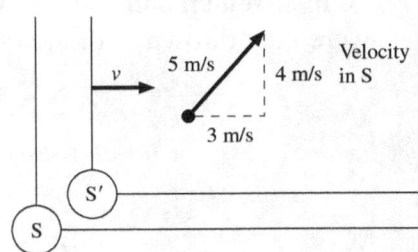

   b. Is there a value of v for which the ball has a minimum speed in S'? If so, what is v? If not, why not?

3. What are the speed and direction of each ball in a reference frame that moves to the right at 2 m/s?

4. Anita is running to the right at 5 m/s. Balls 1 and 2 are thrown toward her at 10 m/s by friends standing on the ground. According to Anita, which ball is moving faster? Or are both speeds the same? Explain.

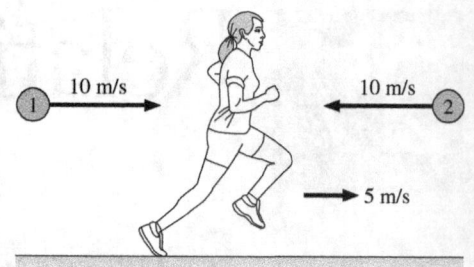

5. Anita is running to the right at 5 m/s. Balls 1 and 2 are thrown toward her by friends standing on the ground. According to Anita, both balls are approaching her at 10 m/s. Which ball was thrown at a faster speed? Or were they thrown with the same speed? Explain.

27.3 Einstein's Principle of Relativity

6. Teenagers Sam and Tom are playing chicken in their
   rockets. As seen from the earth, each is traveling at
   $0.95c$ as he approaches the other. Sam fires a laser
   beam toward Tom.

   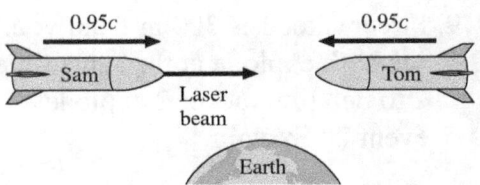

   a. What is the speed of the laser beam relative to Sam?

   b. What is the speed of the laser beam relative to Tom?

27.4 Events and Measurements

7. It is a bitter cold day at the South Pole, so cold that the speed of sound is only 300 m/s. The
   speed of light, as always, is 300 m/μs. A firecracker explodes 600 m away from you.

   a. How long after the explosion until you see the flash of light? _____

   b. How long after the explosion until you hear the sound? _____

   c. Suppose you see the flash at $t = 2.000002$ s. At what time was the explosion? _____

   d. What are the spacetime coordinates for the event "firecracker explodes"? Assume that you
      are at the origin and that the explosion takes place at a position on the positive x-axis.

8. You are at the origin of a coordinate system containing clocks, but you're not sure if the
   clocks have been synchronized. The clocks have reflective faces, allowing you to read them
   by shining light on them. You flash a bright light at the origin at the instant your clock reads
   $t = 2.000000$ s.

   a. At what time will you see the reflection of the light from a clock at $x = 3000$ m?

   b. When you see the clock at $x = 3000$ m, it reads 2.000020 s. Is the clock synchronized with
      your clock at the origin? Explain.

27.5 The Relativity of Simultaneity

9. Firecracker 1 is 300 m from you. Firecracker 2 is 600 m from you in the same direction. You see both explode at the same time. Define event 1 to be "firecracker 1 explodes" and event 2 to be "firecracker 2 explodes." Does event 1 occur before, after, or at the same time as event 2? Explain.

10. Firecrackers 1 and 2 are 600 m apart. You are standing exactly halfway between them. Your lab partner is 300 m on the other side of firecracker 1. You see two flashes of light, from the two explosions, at exactly the same instant of time. Define event 1 to be "firecracker 1 explodes" and event 2 to be "firecracker 2 explodes." According to your lab partner, based on measurements he or she makes, does event 1 occur before, after, or at the same time as event 2? Explain.

11. Can two spatially separated events be simultaneous if they are seen at two different times? If not, why not? If so, give an example.

12. Can two simultaneous events, A and B, at different locations be seen by different people as taking place in a different order, so that one person sees A then B, but another person sees B followed by A? If not, why not? If so, give an example.

13. Two trees are 600 m apart. You are standing exactly halfway between them and your lab partner is at the base of tree 1. Lightning strikes both trees.

   a. Your lab partner, based on measurements he or she makes, determines that the two lightning strikes were simultaneous. What did you see? Did you see the lightning hit tree 1 first, hit tree 2 first, or hit them both at the same instant of time? Explain.

   b. Lightning strikes again. This time your lab partner sees both flashes of light at the same instant of time. What did you see? Did you see the lightning hit tree 1 first, hit tree 2 first, or hit them both at the same instant of time? Explain.

   c. In the scenario of part b, were the lightning strikes simultaneous? Explain.

14. A rocket is traveling from left to right. At the instant it is halfway between two trees, lightning simultaneously (in the rocket's frame) hits both trees.

   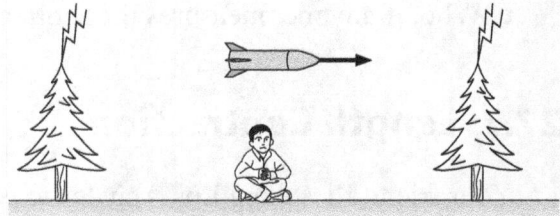

   a. Do the light flashes reach the rocket pilot simultaneously? If not, which reaches him first? Explain.

   b. A student was sitting on the ground halfway between the trees as the rocket passed overhead. According to the student, were the lightning strikes simultaneous? If not, which tree was hit first? Explain.

27.6 Time Dilation

15. Clocks C_1 and C_2 in frame S are synchronized. Clock C' moves at speed v relative to frame S. Clocks C' and C_1 read exactly the same as C' goes past. As C' passes C_2, is the time shown on C' earlier than, later than, or the same as the time shown on C_2? Explain.

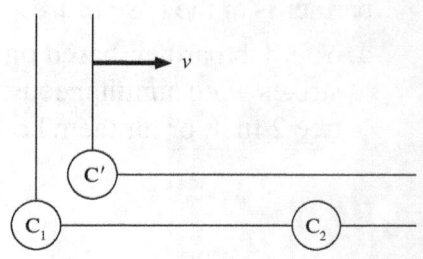

16. Your friend flies from Los Angeles to New York. She carries an accurate stopwatch with her to measure the flight time. You and your assistants on the ground also measure the flight time.

    a. Identify the two events associated with this measurement.

    b. Who, if anyone, measures the proper time? _____

    c. Who, if anyone, measures the shorter flight time? _____

27.7 Length Contraction

17. Your friend flies from Los Angeles to New York. He determines the distance using the tried-and-true $d = vt$. You and your assistants on the ground also measure the distance, using meter sticks and surveying equipment.

    a. Who, if anyone, measures the proper length? _____

    b. Who, if anyone, measures the shorter distance? _____

18. Experimenters in B's reference frame measure $L_A = L_B$. Do experimenters in A's reference frame agree that A and B have the same length? If not, which do they find to be longer? Explain.

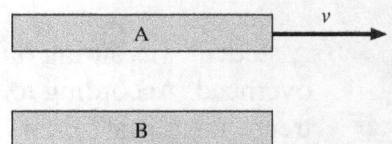

27.8 Velocities of Objects in Special Relativity

19. A rocket travels at speed $0.5c$ relative to the earth.

 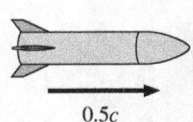

a. The rocket shoots a bullet in the forward direction at speed $0.5c$ relative to the rocket. Is the bullet's speed relative to the earth less than, greater than, or equal to c?

b. The rocket shoots a second bullet in the backward direction at speed $0.5c$ relative to the rocket. In the earth's frame, is the bullet moving right, moving left, or at rest?

27.9 Relativistic Momentum

20. Particle A has half the mass and twice the speed of particle B. Is p_A less than, greater than, or equal to p_B? Explain.

21. Particle A has one-third the mass of particle B. The two particles have equal momenta. Is u_A less than, greater than, or equal to $3u_B$? Explain.

22. Event B occurs at $t_B = 10.0$ μs. An earlier event A, at $t_A = 5.0$ μs, is the cause of B. What is the maximum possible distance that A can be from B?

27.10 Relativistic Energy

23. Can a particle of mass m have total energy less than mc^2? Explain.

24. Consider these 4 particles:

Particle	Rest energy	Total energy
1	A	A
2	B	$2B$
3	$2C$	$4C$
4	$3D$	$5D$

Rank in order, from largest to smallest, the particles' speeds u_1 to u_4.

Order:

Explanation:

28 Quantum Physics

28.1 X Rays and X-Ray Diffraction

1. Use trigonometry to show on the diagram that the path-length difference between the two x rays is $\Delta r = 2d\cos\theta$.

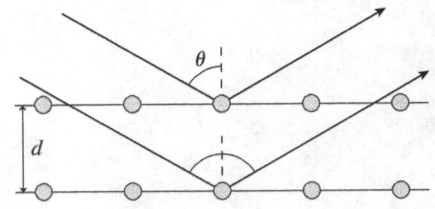

28.2 The Photoelectric Effect

2. a. A negatively charged electroscope can be discharged by shining an ultraviolet light on it. How does this happen?

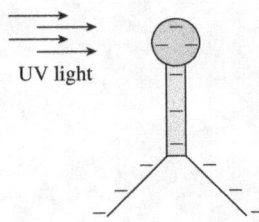

b. You might think that an ultraviolet light shining on an initially uncharged electroscope would cause the electroscope to become positively charged as photoelectrons are emitted. In fact, ultraviolet light has no noticeable effect on an uncharged electroscope. Why not?

3. In the photoelectric effect experiment, a current is measured while light is shining on the cathode. But this does not appear to be a complete circuit, so how can there be a current? Explain.

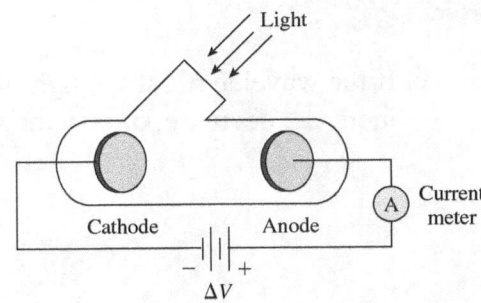

4. The work function of a metal measures:

   i.   The kinetic energy of the electrons in the metal.

   ii. How tightly the electrons are bound within the metal.

   iii. The amount of work done by the metal when it expands.

Which of these (perhaps more than one) are correct? Explain.

5. a. What is the significance of V_{stop}? That is, what have you learned if you measure V_{stop}?
   Note: Don't say that "$-V_{stop}$ is the potential that causes the current to stop." That is merely the definition of V_{stop}. It doesn't say what the *significance* of V_{stop} is.

  b. Why is it surprising that V_{stop} is independent of the light intensity? What would you *expect* V_{stop} to do as the intensity increases? Explain.

  c. If the wavelength of the light in a photoelectric effect experiment is increased, does V_{stop} increase, decrease, or stay the same? Explain.

6. The figure shows a typical current-versus-potential difference graph for a photoelectric effect experiment. On the figure, draw and label graphs for the following three situations:

    i.   The light intensity is increased.
    ii. The light frequency is increased.
    iii. The cathode work function is increased.

    In each case, no other parameters of the experiment are changed.

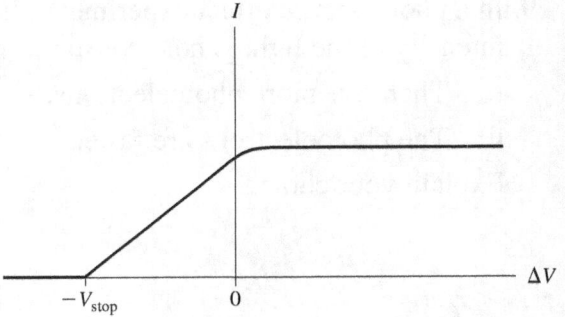

7. The figure shows a typical current-versus-frequency graph for a photoelectric effect experiment. On the figure, draw and label graphs for the following three situations:

    i.   The light intensity is increased.
    ii. The anode-cathode potential difference is increased.
    iii. The cathode work function is increased.

    In each case, no other parameters of the experiment are changed.

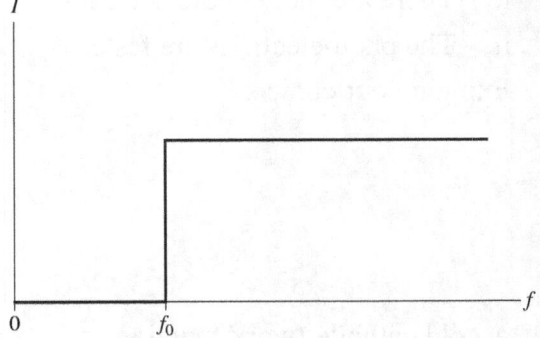

8. The figure shows a typical stopping potential-versus-frequency graph for a photoelectric effect experiment. On the figure, draw and label graphs for the following two situations:

    i.   The light intensity is increased.
    ii. The cathode work function is increased.

    In each case, no other parameters of the experiment are changed.

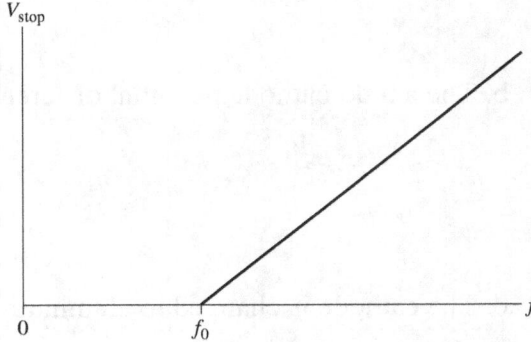

9. In a photoelectric effect experiment, the wavelength of the light is increased while the intensity of the light is held constant. As a result:

   i.   There are more photoelectrons.   iii. Both i and ii.

   ii. The photoelectrons are faster.   iv. Neither i nor ii.

   Explain your choice.

10. In a photoelectric effect experiment, the intensity of light is increased while the wavelength of light is held constant. As a result:

   i.   There are more photoelectrons.   iii. Both i and ii.

   ii. The photoelectrons are faster.   iv. Neither i nor ii.

   Explain your choice.

11. A gold cathode (work function = 5.1 eV) is illuminated with light of wavelength 250 nm. It is found that the photoelectron current is zero when $\Delta V = 0$ V. Would the current change if:

   a. The intensity is doubled?

   b. The anode-cathode potential difference is increased to $\Delta V = 5.5$ V?

   c. The cathode is changed to aluminum (work function = 4.3 eV)?

28.3 Photons

12. The top figure is the *negative* of the photograph of a single-slit diffraction pattern. That is, the darkest areas in the figure were the brightest areas on the screen. This photo was made with an extremely large number of photons.

Suppose the slit is illuminated by an extremely weak light source, so weak that only 1 photon passes through the slit every second. Data are collected for 60 seconds. Draw 60 dots on the empty screen to show how you think the screen might look after 60 photons have been detected.

13. Light of wavelength $\lambda = 1$ μm is emitted from point A. A photon is detected 5 μm away at point B. On the figure, draw the trajectory that a photon follows between points A and B.

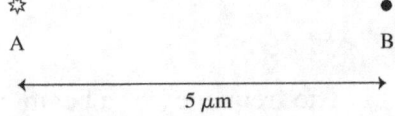

14. Three laser beams have wavelengths $\lambda_1 = 400$ nm, $\lambda_2 = 600$ nm, and $\lambda_3 = 800$ nm. The power of each laser beam is 1 W.

a. Rank in order, from largest to smallest, the photon energies E_1, E_2, and E_3 in these three laser beams.

Order:

Explanation:

b. Rank in order, from largest to smallest, the number of photons per second N_1, N_2, and N_3 delivered by the three laser beams.

Order:

Explanation:

15. When we say that a photon is a "quantum of light," what does that mean? What is quantized?

16. The intensity of a beam of light is increased but the light's frequency is unchanged. As a result:

    i.   The photons travel faster.   iii. The photons are larger.
    ii. Each photon has more energy.   iv.   There are more photons per second.

    Which of these (perhaps more than one) are true? Explain.

17. The frequency of a beam of light is increased but the light's intensity is unchanged. As a result:

    i.   The photons travel faster.   iii. There are fewer photons per second.
    ii. Each photon has more energy.   iv.   There are more photons per second.

    Which of these (perhaps more than one) are true? Explain.

28.4 Matter Waves

18. The figure is a simulation of the electrons detected behind a very narrow double slit. Each bright dot represents one electron. How will this pattern change if the following experimental conditions are changed? Possible changes you should consider include the number of dots and the spacing, width, and positions of the fringes.

    a. The electron-beam intensity is increased.

    b. The electron speed is reduced.

    c. The electrons are replaced by positrons with the same speed. Positrons are antimatter particles that are identical to electrons except that they have a positive charge.

    d. One slit is closed.

19. Very slow neutrons pass through a single, very narrow slit. Use 50 or 60 dots to show how the neutron intensity will appear on a neutron-detector screen behind the slit.

20. Electron 1 is accelerated from rest through a potential difference of 100 V. Electron 2 is accelerated from rest through a potential difference of 200 V. Afterward, which electron has the larger de Broglie wavelength? Explain.

21. An electron and a proton are each accelerated from rest through a potential difference of 100 V. Afterward, which particle has the larger de Broglie wavelength? Explain.

22. Neutron beam 1 has a temperature of 500 K. Neutron beam 2 has a temperature of 5 K. Which neutrons have the larger de Broglie wavelength? Explain.

23. A neutron is shot straight up with an initial speed of 100 m/s. As it rises, does its de Broglie wavelength increase, decrease, or not change? Explain.

24. Double-slit interference of electrons occurs because:
    i.   The electrons passing through the two slits repel each other.
    ii. Electrons collide with each other behind the slits.
    iii. Electrons collide with the edges of the slits.
    iv. Each electron goes through both slits.
    v.   The energy of the electrons is quantized.
    vi. Only certain wavelengths of the electrons fit through the slits.

    Which of these (perhaps more than one) are correct? Explain.

28.5 Energy Is Quantized

25. For the first allowed energy of a particle in a box to be large, should the box be very big or very small? Explain.

26. The smallest allowed energy of a particle in a box is 2.0 eV. What will the smallest energy be if:

    a. The length of the box is doubled?

    b. The mass of the particle is halved?

27. The figure shows the standing de Broglie wave of a particle in a box.

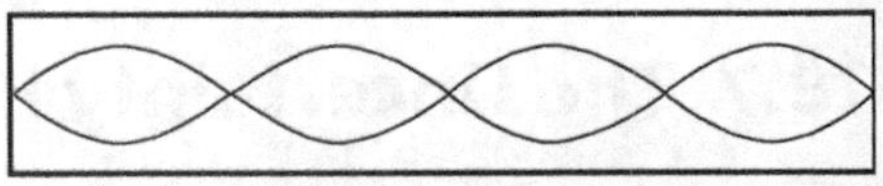

    a. What is the quantum number?

    b. Can you determine from this picture whether the "classical" particle is moving to the right or the left? If so, which is it? If not, why not?

28. A particle in a box of length L_a has $E_1 = 2$ eV. The same particle in a box of length L_b has $E_2 = 50$ eV. What is the ratio L_a/L_b?

28.6 Energy Levels and Quantum Jumps

29. The energy-level diagram is shown for a particle with quantized energy.

   a. On the left, draw an arrow or arrows to show all transitions or quantum jumps in which a particle in the $n = 3$ state absorbs a photon. Below the figure, list E_{photon} for any photon or photons that can be absorbed. Not all blanks may be needed. Don't forget units!

   b. On the right, draw an arrow or arrows to show all transitions or quantum jumps in which a particle in the $n = 3$ state emits a photon. Below the figure, list E_{photon} for any photon or photons that can be emitted. Not all blanks may be needed. Don't forget units!

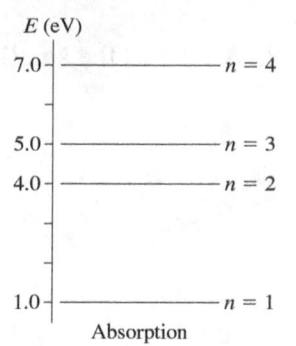 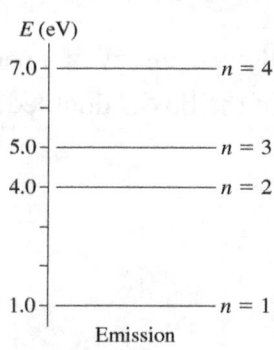

$E_{photon} = $ _____ _____ _____ $E_{photon} = $ _____ _____ _____

   c. Might this quantum system emit a photon with $E_{photon} = 3.0$ eV? If so, how could this happen? If not, why not?

28.7 The Uncertainty Principle

30. A highly collimated beam of electrons passes through a 50-nm-wide slit. An electron detector behind the slit displays a single-slit diffraction pattern with a 2.0-μm-wide central maximum. Assume that the x-axis is parallel to the plane of the slit and the plane of the detector.

   a. As an electron leaves the slit, the uncertainty in its x-position is $\Delta x = $ _____

   b. Does each electron leaving the slit have the same value of p_x? Or do different electrons have different values of p_x? Explain.

   c. Suppose the slit width is narrowed to 30 nm. Will the width of the central maximum then be greater than, less than, or equal to 2.0 μm? _____

   d. With the narrower slit, has Δx increased, decreased, or stayed the same? _____

   e. Has Δp_x increased, decreased, or stayed the same? _____

28.8 Applications and Implications of Quantum Theory

No exercises for this section.

29 Atoms and Molecules

29.1 Spectroscopy

1. The figure shows the emission spectrum of a gas discharge tube.

400 nm   500 nm   600 nm   700 nm

What color would the discharge appear to your eye? Explain.

2. A photograph of an absorption spectrum is a rainbow with black lines. A photograph of an emission spectrum is black with bright colored lines. Why are they different?

29.2 Atoms

3. Suppose you throw a small, hard rubber ball through a tree. The tree has many outer leaves, so you cannot see clearly into the tree. Most of the time your ball passes through the tree and comes out the other side with little or no deflection. On occasion, the ball emerges at a very large angle to your direction of throw. On rare occasions, it even comes straight back toward you. From these observations, what can you conclude about the structure of the tree? Be specific as to how you arrive at these conclusions *from the observations*.

4. Beryllium is the fourth element in the periodic table. Draw pictures similar to Figure 29.8 showing the structure of neutral Be, of Be$^+$, of Be^{++}, and of the negative ion Be$^-$.

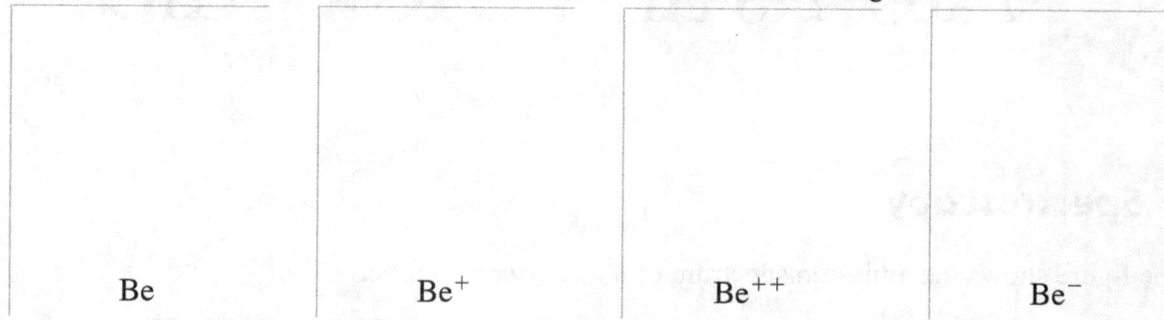

Be   Be$^+$ Be^{++} Be$^-$

5. The element hydrogen has three isotopes. The most common has $A = 1$. A rare form of hydrogen (called *deuterium*) has $A = 2$. An unstable, radioactive form of hydrogen (called *tritium*) has $A = 3$. Draw pictures similar to Figure 29.10 showing the structure of these three isotopes. Show all the electrons, protons, and neutrons of each.

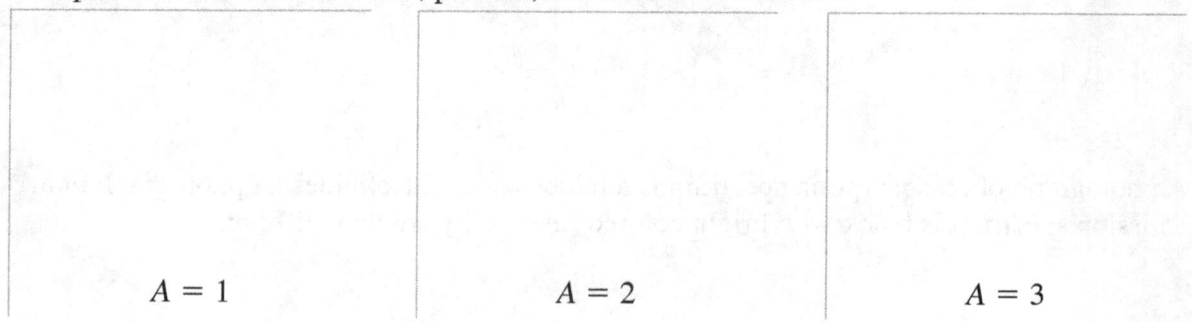

$A = 1$ $A = 2$ $A = 3$

6. Identify the element, the isotope, and the charge state. Give your answer in symbolic form, such as ^{4}He$^+$ or ^{8}Be$^-$.

a.   b.

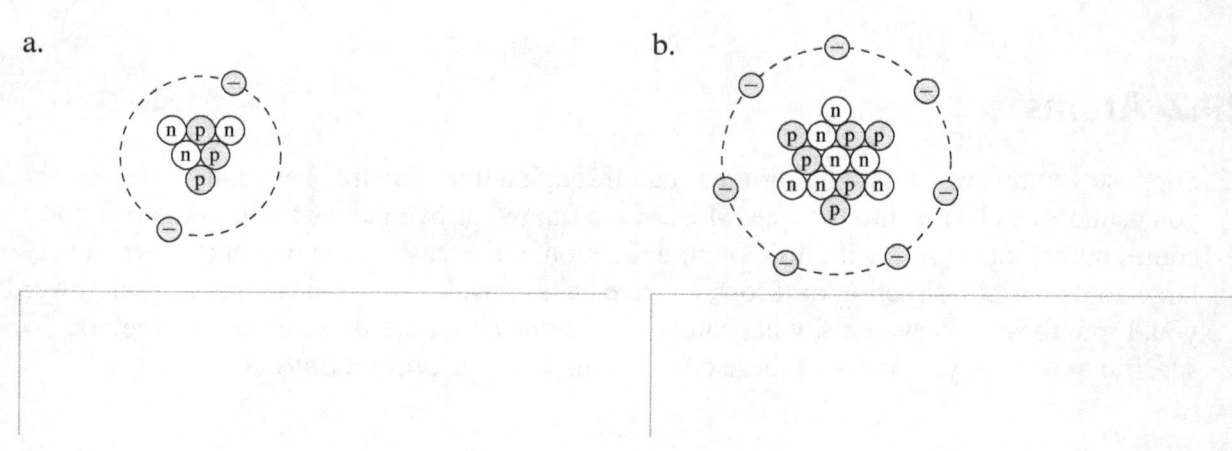

29.3 Bohr's Model of Atomic Quantization

29.4 The Bohr Hydrogen Atom

7. The figure shows a hydrogen atom, with an electron orbiting a proton.

   a. What force or forces act on the electron?

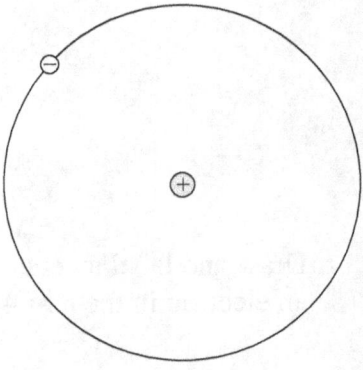

   b. On the figure, draw and label the electron's velocity, acceleration, and force vectors.

8. a. The stationary state of hydrogen shown on the left has quantum number $n =$ _____

   b. On the right, draw the stationary state of the $n - 1$ state.

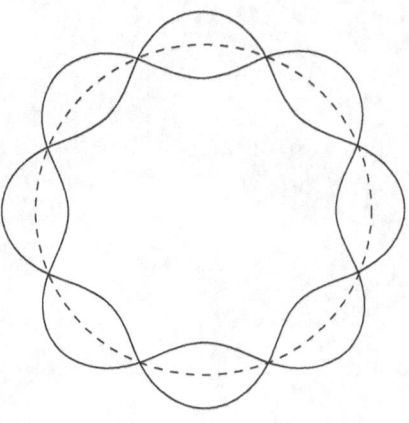 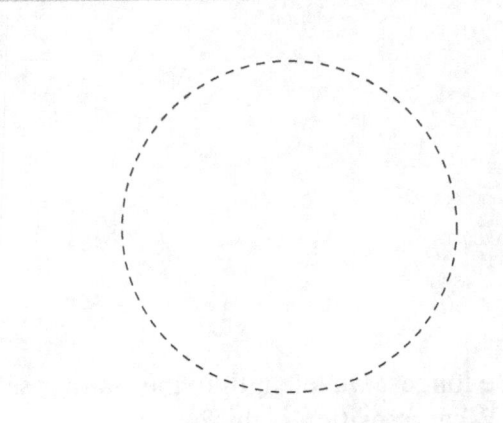

9. Bohr did not include the gravitational force in his analysis of the hydrogen atom. Is this one of the reasons that his model of the hydrogen atom had only limited success? Explain.

10. Why is there no stationary state of hydrogen with $E = -9$ eV?

11. Draw and label an energy level diagram for hydrogen. On it, show all the transitions by which an electron in the $n = 4$ state could emit a photon.

12. The longest wavelength in the Balmer series is 656 nm.
    a. What transition is this?

    b. If light of this wavelength shines on a container of hydrogen atoms, will the light be absorbed? Why or why not?

29.5 The Quantum-Mechanical Hydrogen Atom

13. List all possible states of a hydrogen atom that have $E = -1.51$ eV.

    $$n \qquad\qquad l \qquad\qquad m$$

14. What are the n and l values of the following states of a hydrogen atom?

    State $= 4d$ $n = $ _____ $l = $ _____

    State $= 5f$ $n = $ _____ $l = $ _____

    State $= 6s$ $n = $ _____ $l = $ _____

15. How would you label the hydrogen-atom states with the following quantum numbers?

    $(n, l, m) = (4, 3, 0)$ Label $= $ _____

    $(n, l, m) = (3, 2, 1)$ Label $= $ _____

    $(n, l, m) = (3, 2, -1)$ Label $= $ _____

16. Consider the two hydrogen-atom states $5d$ and $4f$. Which has the higher energy? Explain.

29.6 Multielectron Atoms

17. Do the following figures represent a possible electron configuration of an element? If so:
    i. Identify the element, and
    ii. Determine if this is the ground state or an excited state.
    If not, why not?

a.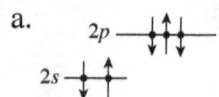

b.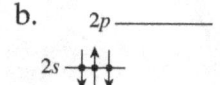

c.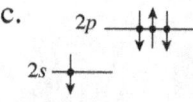

18. Do the following electron configurations represent a possible state of an element? If so:
    i. Identify the element, and
    ii. Determine if this is the ground state or an excited state.
    If not, why not?

    a. $1s^2 2s^2 2p^6 3s^2$

    b. $1s^2 2s^2 2p^7 3s$

    c. $1s^2 2s^2 2p^6 3s 3p^2$

19. Why is the section of the periodic table labeled "transition elements" exactly 10 elements wide in all rows?

29.7 Excited States and Spectra

20. The figure shows the energy levels of a hypothetical atom.

    a. What is the atom's ionization energy?

    b. In the space below, draw the energy-level diagram as it would appear if the ground state were chosen as the zero of energy. Label each level and the ionization limit with the appropriate energy.

eV

-------------------------------- 0
                        7d   −1
              7p   −2

      7s   −4

                   6p   −7

   6s   −9
Ground state

21. The figure shows the energy levels of a hypothetical atom.

    a. What *minimum* kinetic energy (in eV) must an electron have to collisionally excite this atom and cause the emission of a 620 nm photon? Explain.

    b. Can an electron with $K = 6$ eV cause the emission of 620 nm light? If so, what is the final kinetic energy of the electron? If not, why not?

    c. Can a 6 eV photon cause the emission of 620 nm light from this atom? Why or why not?

    d. Can 7 eV photons cause the emission of 620 nm light from these atoms? Why or why not?

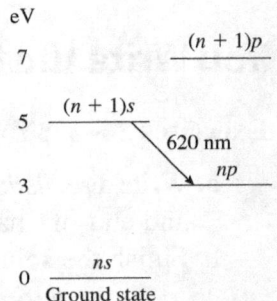

eV

7                    $(n + 1)p$

5        $(n + 1)s$
                      620 nm
3   np

0   ns
       Ground state

29.8 Molecules

22. Laundry detergent manufacturers sometimes use fluorescent chemicals to make clothes appear whiter in the presence of small amounts of UV light. What role does UV light (which is not directly visible) play in making these "whites appear whiter"?

29.9 Stimulated Emission and Lasers

23. A photon with energy 2.0 eV is an incident on an atom in the p-state. Does the atom undergo an absorption transition, a stimulated emission transition, or neither? Explain.

You Write the Problem!

Exercises 24–25: You are given the equation that is used to solve a problem. For each of these:

    a. Write a *realistic* physics problem for which this is the correct equation. Look at worked examples and end-of-chapter problems in the textbook to see what realistic physics problems are like.

    b. Finish the solution of the problem.

24. $496 \text{ nm} = \dfrac{1240 \text{ eV} \cdot \text{nm}}{\Delta E_{\text{atom}}}$

25. $0.476 \text{ nm} = n^2 (0.0529 \text{ nm})$

$$E_n = -\frac{13.60 \text{ eV}}{n^2}$$

30 Nuclear Physics

30.1 Nuclear Structure

1. Consider the atoms ^{16}O, ^{18}O, ^{18}F, ^{18}Ne, and ^{20}Ne. Some of the questions about these atoms may have more than one answer. Give all answers that apply.

   a. Which atoms are isotopes? _____

   b. Which atoms have the same number of nucleons? _____

   c. Which atoms have the same chemical properties? _____

   d. Which atoms have the same number of neutrons? _____

   e. Which atoms have the same number of valence electrons? _____

2. Nuclear physics calculations often involve finding the *difference* between two almost equal numbers, which requires careful attention to significant figures. Suppose you need to know the *difference* between the neutron mass and the proton mass to two significant figures. To how many significant figures must you know the masses themselves if

   a. the masses are given in atomic masses units (u)? _____

   b. the masses are given in MeV/c^2? _____

30.2 Nuclear Stability

3. a. Is the total binding energy of a nucleus with $A = 200$ more than, less than, or equal to the binding energy of a nucleus with $A = 60$? Explain.

   b. Is a nucleus with $A = 200$ more tightly bound, less tightly bound, or bound equally tightly as a nucleus with $A = 60$? Explain.

4. a. Is there a ^{30}Li ($Z = 3$) nucleus? If so, is it stable or radioactive? If not, why not?

b. Is there a ^{230}U ($Z = 92$) nucleus? If so, is it stable or radioactive? If not, why not?

5. Rounding slightly, the nucleus ^{3}He has a binding energy of 2.5 MeV/nucleon and the nucleus ^{6}Li has a binding energy of 5 MeV/nucleon.

   a. What is the total binding energy of ^{3}He? _____

   b. What is the total binding energy of ^{6}Li? _____

   c. Is it energetically possible for two ^{3}He nuclei to join or fuse together into a ^{6}Li nucleus? Explain.

   d. Is it energetically possible for a ^{6}Li nucleus to split or fission into two ^{3}He nuclei? Explain.

30.3 Forces and Energy in the Nucleus

6. Draw energy-level diagrams showing the nucleons in ^{6}Li and ^{7}Li.

   a.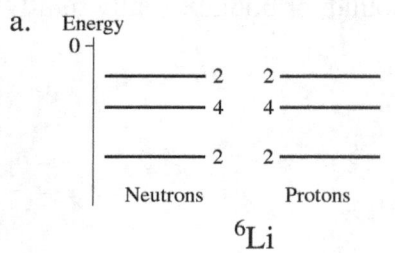

   b.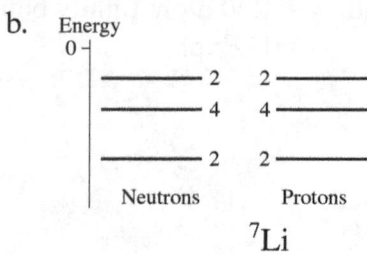

30.4 Radiation and Radioactivity

7. Identify the unknown X in the following decays:

    a. ^{222}Rn ($Z = 86$) → ^{218}Po ($Z = 84$) + X   X = _____

    b. ^{228}Ra ($Z = 88$) → ^{228}Ac ($Z = 89$) + X   X = _____

    c. ^{140}Xe ($Z = 54$) → ^{140}Cs ($Z = 55$) + X   X = _____

    d. ^{64}Cu ($Z = 29$) → ^{64}Ni ($Z = 28$) + X   X = _____

    e. ^{18}F ($Z = 9$) + X → ^{18}O ($Z = 8$)   X = _____

8. Are the following decays possible? If not, why not?

    a. ^{232}Th ($Z = 90$) → ^{236}U ($Z = 92$) + α

    b. ^{238}Pu ($Z = 94$) → ^{236}U ($Z = 92$) + α

    c. ^{11}Na ($Z = 11$) → ^{11}Na ($Z = 11$) + γ

    d. ^{33}P ($Z = 15$) → ^{32}S ($Z = 16$) + e^-

9. Part of the ^{236}U decay series is shown.
    a. Complete the labeling of atomic numbers and elements on the bottom edge.
    b. Label each arrow to show the type of decay.

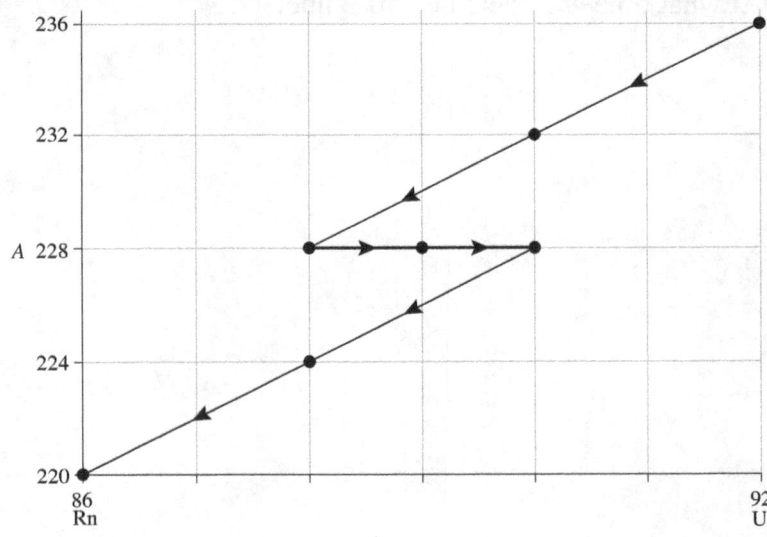

30.5 Nuclear Decay and Half-Lives

10. The half-life of the radioactive nucleus A is 6 hours.

   a. Plot the number of A nuclei remaining as a function of time on the axes below, assuming there were 1,000,000 nuclei at time $t = 0$.

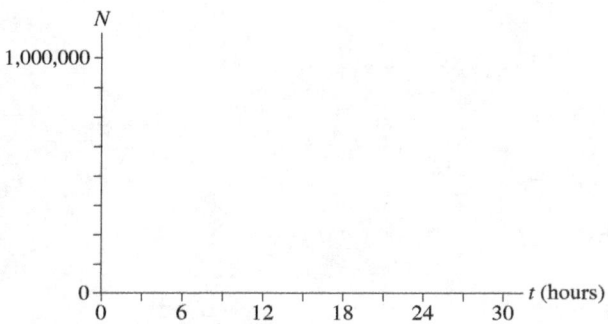

   b. How many A nuclei remain after one day? _____

11. What is the half-life of this nucleus?

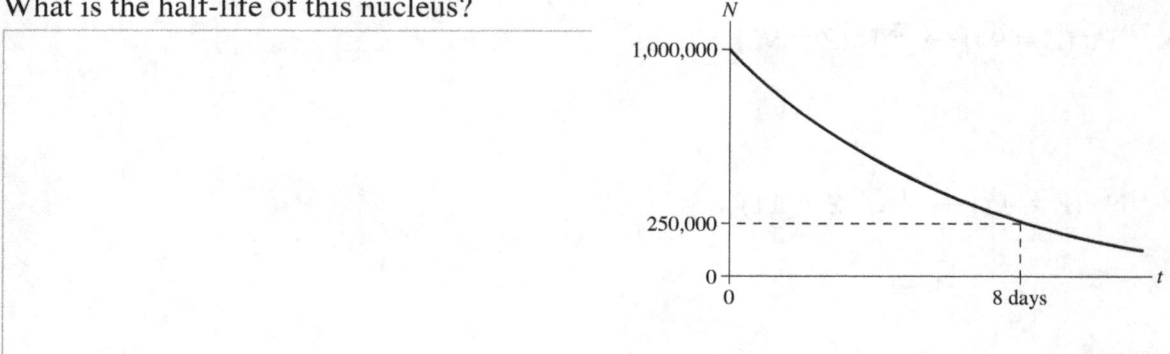

12. A radioactive sample B has a half-life of 10 s. 10,000 B nuclei are present at $t = 20$ s.

   a. How many B nuclei were there at $t = 0$ s? _____

   b. How many B nuclei will there be at $t = 40$ s? _____

13. Nucleus A decays into nucleus B with a half-life of 10 s. At $t = 0$ s, there are 1000 A nuclei and no B nuclei. At what time will there be 750 B nuclei?

30.6 Medical Applications of Nuclear Physics

14. What is the difference, or is there a difference, between radiation *dose* and *dose equivalent*?

15. An apple is irradiated for one hour by an intense beam of alpha radiation. Afterward, is the apple radioactive? Why or why not?

16. Four patients receive radiation therapy.

Patient	Radiation	Dose (Gy)	RBE
A	X rays	0.01	1
B	Alpha	0.01	20
C	Beta	0.05	1
D	Protons	0.02	5

a. Rank in order, from largest to smallest, the energy absorbed by each patient.

Order:

Explanation:

b. Rank in order, from largest to smallest, the biological damage incurred by each patient.

Order:

Explanation:

30.7 The Ultimate Building Blocks of Matter

17. A subatomic particle called the *pion* has the structure $u\bar{d}$; that is, it is made of an up quark and an antidown quark.

    a. What is the charge of a pion? Explain.

    b. What is the structure of an antipion? What is its charge?

18. One family of subatomic particles made of three up (u) and/or down (d) quarks is the Δ family. The Δ^{++} particle has the interesting property that its charge is $+2e$. What is the quark structure of the Δ^{++}? Explain.

You Write the Problem!

Exercises 19–20: You are given the equation that is used to solve a problem. For each of these:

   a. Write a *realistic* physics problem for which this is the correct equation. Look at worked examples and end-of-chapter problems in the textbook to see what realistic physics problems are like.

   b. Finish the solution of the problem.

19. $3.75 \times 10^{12} = 6.00 \times 10^{13} \left(\dfrac{1}{2}\right)^{(2.5\ \text{min})/t_{1/2}}$

20. $3.0\ \text{Sv} = \dfrac{\text{dose}}{0.035\ \text{kg}} \times 5$

DYNAMICS WORKSHEET Name _____ Problem _____

PREPARE

- List knowns. Identify what you're trying to find.
- Identify forces and draw a free-body diagram.

- Draw a pictorial representation for problems with motion: Show important points in the motion, establish a coordinate system, define symbols, draw a motion diagram.

Known

Find

SOLVE

Start with Newton's first or second law in component form, adding other information as needed to solve the problem.

ASSESS

Have you answered the question?

Do you have correct units, signs, and significant figures?

Is your answer reasonable?

DYNAMICS WORKSHEET Name _____ Problem _____

PREPARE

- List knowns. Identify what you're trying to find.
- Identify forces and draw a free-body diagram.

- Draw a pictorial representation for problems with motion: Show important points in the motion, establish a coordinate system, define symbols, draw a motion diagram.

Known

Find

SOLVE

Start with Newton's first or second law in component form, adding other information as needed to solve the problem.

ASSESS

Have you answered the question?
Do you have correct units, signs, and significant figures?
Is your answer reasonable?

MOMENTUM WORKSHEET Name _____ Problem _____

PREPARE
- Identify the system and any external forces.
- Sketch "before and after."
- Define coordinates.
- List knowns. Identify what you're trying to find.

Known

Find

- Is momentum conserved? _____

SOLVE
Start with conservation of momentum or the impulse-momentum theorem, using Newton's laws or kinematics as needed.

ASSESS
Have you answered the question?
Do you have correct units, signs, and significant figures?
Is your answer reasonable?

MOMENTUM WORKSHEET Name _____ Problem _____

PREPARE
- Identify the system and any external forces.
- Sketch "before and after."
- Define coordinates.
- List knowns. Identify what you're trying to find.

Known

Find

- Is momentum conserved?

SOLVE
Start with conservation of momentum or the impulse-momentum theorem, using Newton's laws or kinematics as needed.

ASSESS
Have you answered the question?
Do you have correct units, signs, and significant figures?
Is your answer reasonable?

ENERGY WORKSHEET

Name _____ Problem _____

PREPARE

- Identify the system.
- Sketch "before and after."
- Define coordinates.
- List knowns. Identify what you're trying to find.

Known

Find

- Is the system isolated? _____   Which energies change? _____

SOLVE

Start with conservation of energy, adding other information and techniques as needed to solve the problem.

ASSESS

Have you answered the question?
Do you have correct units, signs, and significant figures?
Is your answer reasonable?

ENERGY WORKSHEET

Name _____ Problem _____

PREPARE
- Identify the system.
- Sketch "before and after."
- Define coordinates.
- List knowns. Identify what you're trying to find.

Known

Find

- Is the system isolated? _____ Which energies change? _____

SOLVE

Start with conservation of energy, adding other information and techniques as needed to solve the problem.

ASSESS

Have you answered the question?
Do you have correct units, signs, and significant figures?
Is your answer reasonable?